The Institute of Mathematics
and its Applications
Conference Series

The Institute of Mathematics
and its Applications
Conference Series

Previous volumes in this series were published by
Academic Press to whom all enquiries should be addressed.
Forthcoming volumes will be published by
Oxford University Press throughout the world.

NEW SERIES
1. *Supercomputers and parallel computation* Edited by D. J. Paddon
2. *The mathematical basis of finite element methods*
   Edited by David F. Griffiths
3. *Multigrid methods for integral and differential equations*
   Edited by D. J. Paddon and H. Holstein
4. *Turbulence and diffusion in stable environments* Edited by J. C. R. Hunt

Turbulence and diffusion in stable environments

Based on the proceedings of a conference on Models of Turbulence and Diffusion in Stably Stratified Regions of the Natural Environment organized by The Institute of Mathematics and its Applications and held in Cambridge, March 1983

Edited by

J. C. R. HUNT
University of Cambridge

CLARENDON PRESS · OXFORD · 1985

Oxford University Press, Walton Street, Oxford OX2 6DP

Oxford New York Toronto
Delhi Bombay Calcutta Madras Karachi
Kuala Lumpur Singapore Hong Kong Tokyo
Nairobi Dar es Salaam Cape Town
Melbourne Auckland

and associated companies in
Beirut Berlin Ibadan Nicosia

Oxford is a trademark of Oxford University Press

Published in the United States
by Oxford University Press, New York

British Library Cataloguing in Publication Data
Models of Turbulence and Diffusion in Stably Stratified
Regions of the Natural Environment (Conference:
1983: Cambridge)
Turbulence and diffusion in stable environments:
based on the proceedings of a conference on Models of
Turbulence and Diffusion in Stably Stratified Regions
of the Natural Environment organized by the Institute
of Mathematics and its Applications and held in
Cambridge, March 1983. − (Institute of Mathematics and
its Applications conference series. New series; 4)
1. Atmospheric turbulence − Mathematical models
2. Diffusion − Mathematical models
I. Title   II. Hunt, J.C.R.   III. Series
551.5'17   QC880.4.T8
ISBN 0-19-853604-6

Printed in Great Britain by St Edmundsbury Press,
Bury St Edmunds, Suffolk

PREFACE

This is a volume of papers presented to a conference held in Cambridge in March 1983 on 'Models of turbulence and diffusion in stably stratified regions of the natural environment', and organised by the Institute of Mathematics and its Applications.

The subject of the conference needs a brief explanation. The environment is generally in a state of motion (even down to the bottom of the oceans), which is usually unsteady and varying from one place to another. These variations in the motion or eddies, which range in scale from thousands of kilometres (e.g., atmospheric cyclones) to millimetres, provide the means for the oceanic and atmospheric environment to mix and diffuse both substances and heat. These eddies also play a crucial <u>dynamical</u> role in controlling the general motion of the atmosphere and oceans. Below scales of about 1km, they are collectively referred to as turbulence.

In the regions of the world where major concentrations of populations and industry discharge large volumes of polluted substances in to the atmosphere, rivers and oceans, the environment is only able to disperse the pollutants to tolerable levels because of the combined action of the turbulence and the larger-scale motions. This continues to be one of the most important reasons why turbulence and diffusion have to be studied, and as far as possible 'modelled'. Quantitative mathematical modelling of the environment is an essential part of effective planning for the utilisation and improvement of the environment. (For an excellent account of the cleaning of the Thames see the article by M.J. Lighthill, Newer Uses of Mathematics, Penguin 1978).

The turbulence in the environment has many different manifestations which are familiar in the atmosphere; the large-scale 'thermals' on sunny days in which gliders spiral upwards and which form white cumulus clouds, violent gusting eddies on a windy day, and at the other extreme a steady late-evening breeze. Similarly distinct kinds of turbulent motions are found in rivers, estuaries and oceans. These different kinds

of motions are found when vertical variation of density in the
environment is unstable, neutrally stratified or stable,
respectively.

At our conference we concentrated on the particular type
of turbulent flow found in the latter stably stratified
environment. The environment is said to be 'stably stratified'
if one finds that a small volume of air (in the atmosphere) or
water (in the aquatic environment) which is displaced upwards
or downwards tends to return to its initial level when it is
released. This occurs wherever the density of the environment
decreases with height rapidly enough, or the temperature
increases rapidly enough. You usually find such conditions in
the atmosphere in cool air near the ground at night, and
most of the time above about 1km. Except when the solar
radiation and wind over the surface are strong enough, the
oceans are mainly stably stratified. Estuaries and rivers are
less likely to be stably stratified, because their currents
are often strong enough to reduce density gradients.

Because of the tendency of vertical motions to be suppressed
by a stable stratification, turbulent motions have special
characteristics in these conditions, namely: vertical
diffusion and the larger scales of vertical motion tend to be
reduced; in shear flows horizontal as well as vertical velocity
fluctuations are reduced; wave motions may be generated by
the turbulence; layers with high density gradients develop
across which the turbulence and the mean velocity change very
rapidly. These effects are found (not usually simultaneously)
in stably stratified flow in the atmosphere and aquatic
environments and can be reproduced and studied in the
laboratory.

Many of the pioneers in turbulence research have made
contributions to the study of turbulence in stably stratified
flows, notably G.I. Taylor, L. Prandtl, T. von Karman,
A.S. Monin, A.M. Obukhov and L.F. Richardson. The Richardson
number, in its many guises, is still regarded as the key
dimensionless parameter defining the effect of stratification
on turbulence. (I have special pleasure recalling Richardson's
contribution to the subject because of the two holidays I and
my brother spent with him and his wife (my great aunt Dolly)
on the Holy Loch in Scotland.) A fine account of Richardson's
life and work has recently been written by the meteorologist
Oliver Ashford (Adam Hilger 1984).

For newcomers to the subject on turbulence and diffusion in
stably stratified flows, two excellent texts are those by
J.S. Turner (Buoyancy Effects in Fluids, Cambridge University

Press, 1973) and G. Csanady (Turbulent Diffusion in the Environment, Reidel, 1973). Some papers in a previous IMA conference are also relevant (Mathematical Modelling of Turbulent Diffusion in the Environment (Ed. C.J. Harris), Academic Press, 1979).

The conference was held because of some of the exciting new developments in this aspect of turbulence research, which have been made possible by new experimental facilities, greater computer power and new theoretical concepts, and because of the many practical uses of this research, particularly in the prediction of dispersion of unwanted products in the natural environment. The papers of this conference reflect many of these new developments and also new controversies.

The first group of papers are on various specific aspects of the interaction of turbulence with a stable density gradient. *R.E. Britter* describes how the decay of turbulence in the absence of a shear is not much affected by stable stratification, but that the rate of vertical diffusion of a pollutant from a point source is strongly reduced to almost zero. He shows how this is related to the reduction in the diffusion coefficient for the vertical flux of density or temperature as the stratification increases. Where there is sufficient energy at frequencies less than the natural frequency of internal motions (the Brunt-Vaisala frequency), significant wave motions can appear in turbulent flows. *E.J. Hopfinger* and *F.K. Browand's* laboratory experiments show the emergence of wave motions from a turbulence generated in a tank by an oscillating grid; they demonstrate how wave motions can, in some circumstances, be more effective in transporting energy and momentum than turbulence can. This transition between turbulence and waves, and the transport of energy by waves, can also be found where there is a sudden change in the stratification at a level where turbulence in a region of a low density gradient adjoins a region with a high density gradient; a topic analysed in a theoretical paper by *D.J. Carruthers and J.C.R. Hunt*. Such transitions are rather widespread in the atmosphere and the oceans.

Another kind of thin transition layer is found at the top of a dense volume of fluid moving into lighter fluid (such as a dense vapour cloud spreading along the ground in the atmosphere), or equivalently at the bottom of a light or warmer layer of fluid flowing above denser or cooler fluid (such as warm water discharged from a power station spreading over the surface of the sea).

These are just two examples of the many practical problems where the transport of heat or pollutants through such layers is the critical factor in determining the temperature or the pollutant concentration. A detailed experimental investigation of this layer is presented in the paper by *N.H. Thomas and J.E. Simpson,* particularly the effects on the flux through the layer of turbulence outside it. This study is one of several in this volume that show how useful quantitative models can be developed by combining physical and dimensional arguments based on experiments, even when a full theoretical or computational study is not possible.

The next two papers are on the general mechanics and modelling of turbulence and diffusion in stably stratified flows; *R.S. Scorer* describes turbulence and its hierarchy of scales of motion in terms of the dynamics of its vorticity, and how the vorticity is changed in the presence of stable stratification. It is a stimulating overview, particularly for readers already familiar with other theoretical models in the literature. *W. Rodi* presents a review of modelling horizontal shear flows with stable stratification when the flux of heat (or density) and the shear stress is proportional to the local mean temperature gradient and mean velocity gradient, respectively, and when the diffusivity of heat and the eddy viscosity can be related to the local kinetic energy of the turbulence (k) and the local rate of dissipation of kinetic energy (ε). A good deal of controversy surrounds the model equations for k and ε in stably stratified flows. (For example, does stable stratification increase or decrease ε?) So, at this stage of our understanding, any models have to be closely compared with experimental results, a nice feature of this review.

Some papers presented at the conference have not been written up for these proceedings, but abstracts are included in this volume. *J.C.R. Hunt* gave a brief review of recent ideas and literature; the reference list is published with the abstract. *B.E. Launder* presented some important ideas about the influences of density fluctuations on the turbulence in stably stratified flows. The results have largely been published elsewhere and some references are given with the abstract.

The second group of papers describes studies of particular environmental flows. *F. Nieuwstadt* presents a new analysis of the mean flow and turbulence in the stably stratified atmospheric boundary layer (S.B.L.) over very flat terrain (using the same 'flux-gradient' relations for the fluxes of heat and momentum reviewed by Rodi). A surprisingly simple structure is predicted which agrees well with the measurements

made on the new 200m meteorological tower at Cabauw in the
Netherlands.

The next two papers on the SBL present measurements and
new concepts about turbulence and wave motion. *S.E. Larsen,
H.R. Olesen and J. Højstrup* argue that in the SBL the
turbulence has a quite different structure at small and large
scales, the three-dimensional small scale spectrum being the
same <u>form</u> as in neutral conditions, but the amplitude is
determined by the surface fluxes of heat and surface shear
stress. But at large scales the <u>structure</u> is strongly affected
by stratification, for example for the horizontal components
the power spectrum is proportional to k^{-3}, where k is the wave
number. *J.C. King* has developed a computational model of
two-dimensional large-scale eddies generated by the character-
istically large velocity gradients to be found in the nocturnal
stable boundary layer. Because of the stratification these
eddies have a wave-like structure, but their amplitude is
strongly dependent on the trapping of the waves within these
stratified shear layers. Experimental evidence for the
importance of trapped internal waves in coastal waters is
provided in the brief paper by *M.F. Lavin and T.A. Sherwin*;
there the waves are often associated with regions of rapidly
changing conditions such as warming, wind stress, or the
presence of fronts.

K.J. Richards' paper shows that many features of stratified
turbulence in the atmospheric and oceanic surface layers are
also present in the boundary layer on the ocean bottom, the
'Benthic' boundary layer. But features that are special to
this boundary layer are the role of synoptic scale eddies in
controlling the height of the layer and the presence of fronts
with significant density differences. These boundary layers
are now being studied intensively because of possible plans
to exploit the ocean flow for disposal of solid waste and
exploitation of mineral resources. Also included in this
section are abstracts of all the papers presented by *S.J.
Caughey* on measurements in the stable atmospheric boundary
layer and by *J. McGuirk, A. Ghobadian, A.J.H. Goddard and
A.D. Gosman,* on modelling wakes of structures in the stable
boundary layer.

The third group of papers is devoted to studies on the
dispersion in the natural environment. *P.C. Chatwin* reviews
some of the basic ideas about modelling the dispersion of dense
gas in the atmosphere as a moving 'box', or control surface,
an effective if slightly artificial concept. He particularly
considers the effect of the velocity gradient or shear on the
control surface, and how it is important to be consistent in

modelling the various phases in the evolution of a dense cloud.
C.D. Jones and D.J. Ride present some recent data on the
dispersion of smoke in the atmosphere as an illuminating
example of turbulent diffusion and of the scales over which
large scales of turbulence are correlated. *J.H. Pickles* and
I.R. Rodgers describe measurements of the evolution of the surface layer of warm water that is discharged from power stations
in coastal waters. Clear evidence is shown how the stable
stratification reduces the vertical flux of heat; a process
that they show can be modelled by means of a suitable eddy
diffusivity (the basic theme of Rodi's paper).

Finally, *R.E. Lewis* presents some important questions about
the mechanisms determining periods of intense mixing in
estuaries, based on his field measurements. Are they more
associated with internal wave or motion than with the arrival
of the salt wedge? Perhaps it is appropriate that the final
paper suggests some further important questions to be studied.

J.C.R. Hunt
University of Cambridge

CONTENTS

List of Contributors   xiii

1. Basic studies and modelling of turbulence and
    diffusion in stratified flows

Diffusion and decay in stably-stratified turbulent flows   3
by R.E. Britter

The inhibition of vertical turbulent scale by stable   15
stratification by F.K. Browand and E.J. Hopfinger

Turbulence and wave motions near an interface between   29
a turbulent region and a stably stratified layer by
D.J. Carruthers and J.C.R. Hunt

Mixing of gravity currents in turbulent surroundings: 61
laboratory studies and modelling implications by
N.H. Thomas and J.E. Simpson

A vivid mechanical picture of turbulence by R.S. Scorer   97

Calculation of stably stratified shear-layer flows   111
with a buoyancy - extended K-ε turbulence model by
W. Rodi

ABSTRACTS

Understanding and modelling turbulence in stably   141
stratified flows by considering displacements and
mixing of fluid elements by J.C.R. Hunt

The turbulence modelling of variable density flows - A   145
mixed-weighted decomposition by B.E. Launder

2. Turbulent stratified flows in the environment

A model for the stationary, stable boundary layer by   149
F.T.M. Nieuwstadt

Parameterization of the low frequency part of spectra   181
of horizontal velocity components in the stable surface
boundary layer by S.E. Larsen, H.R. Olesen and
J. Højstrup

Modelling the development of large eddies in the stable   205
atmospheric boundary layer by J.C. King

Stratification and internal waves in the western Irish   225
Sea by M.F. Lavin and T.J. Sherwin

The benthic boundary layer by K.J. Richards   237

ABSTRACTS

Turbulence and waves in stable layers by S.J. Caughey   253

Calculation of the development of three-dimensional   255
wake flows in a stably stratified environment by
J. McGuirk, A. Ghobadian, A.J.H. Goddard and A.D. Gosman

3. Dispersion in stratified flow in the environment

Towards a box model of all stages of heavy gas cloud   259
dispersion by P.C. Chatwin

Multiple smoke flume trials - The Chemical Defence   293
Establishment by C.D. Jones and D.J. Ride

Vertical heat transport in a cooling water plume by   297
J.H. Pickles and I.R. Rodgers

Intense mixing periods in an estuary by R.E. Lewis   315

LIST OF CONTRIBUTORS

R.W. Britter; *Department of Engineering, University of Cambridge, Trumpington Street, Cambridge CB2 1PZ.*

F.K. Browand; *Institute of Aerospace Engineering, University of Southern California, University Park, Los Angeles, California, 90089-1454, USA.*

D.J. Carruthers; *Department of Applied Mathematics and Theoretical Physics, University of Cambridge, Silver Street, Cambridge CB3 9EW.*

S.J. Caughey; *Principal Meteorological Officer, Meteorological Office, Belfast (Aldergrove) Airport, Belfast BT29 4AB.*

P.C. Chatwin; *Department of Mathematics, Brunel University, Uxbridge, Middlesex UB8 3PH.*

A.J.H. Goddard; *Department of Mechanical Engineering, Imperial College of Science and Technology, Exhibition Road, London SW7 2BX.*

A.D. Gosman; *Department of Mechanical Engineering, Imperial College of Science and Technology, Exhibition Road, London SW7 2BX.*

J. Højstrup; *Risø National Laboratory, Postbox 49, DK-4000 Roskilde, Denmark.*

E.J. Hopfinger; *Institut de Mecanique (I.M.H.), B.P. 53 X, 38041 Grenoble Cedex, France.*

J.C.R. Hunt; *Department of Applied Mathematics and Theoretical Physics, University of Cambridge, Silver Street, Cambridge CB3 9EW.*

C.D. Jones; *ATRD, CDE Porton Down, Salisbury, Wiltshire.*

J.C. King; *Meteorological Research Unit, Royal Air Force, Cardington, Bedford MK42 0TH.*

S.E. Larsen; *Risø National Laboratory, Postbox 49, DK-4000 Roskilde, Denmark.*

B.E. Launder; *Department of Mechanical Engineering, UMIST, P.O. Box 88, Manchester M60 1QD.*

M. Lavin; *Marine Science Laboratory, Menai Bridge, Anglesey, Gwynedd.*

R.W. Lewis; *The Brixham Laboratory, ICL Plc., Freshwater Quarry, Brixham, South Devon.*

J. McGuirk; *Department of Mechanical Engineering, Imperial College of Science and Technology, Exhibition Road, London SW7 2BX.*

F.T.M. Nieuwstadt; *Royal Netherlands Meteorological Institute, p.p. box 201, 3720 AE de bilt, The Netherlands.*

H.R. Olesen; *Institute of Mathematical Statistics and Operational Research, Technical University of Denmark, DK-2800 Lyngby, Denmark.*

J.H. Pickles; *Technology Planning and Research Division, CERL, Kelvin Avenue, Leatherhead, Surrey KT22 7SE.*

K.J. Richards; *Institute of Oceanographic Sciences, Brook Road, Wormley, Godalming, Surrey GU8 5UB.*

D.J. Ride; *ATRD, CDE Porton Down, Salisbury, Wiltshire.*

I.R. Rodgers; *Technology Planning and Research Division, CERL, Kelvin Avenue, Leatherhead, Surrey, KT22 7SE.*

W. Rodi; *Institut für Hydromechanik, Universität Karlsruhe, D-7500 Karlsruhe 1, Kaiserstrasse 12, Postfach 6380, West Germany.*

R.S. Scorer; *Department of Theoretical Mechanics, Imperial College of Science and Technology, London.*

T.J. Sherwin; *Marine Science Laboratory, Menai Bridge, Anglesey, Gwynedd.*

J.E. Simpson; *Department of Applied Mathematics and Theoretical Physics, University of Cambridge, Silver Street, Cambridge CB3 9EW.*

N.H. Thomas; *Department of Chemical Engineering, University of Birmingham, P.O. Box 363, Edgbaston, Birmingham B15 2TT.*

ACKNOWLEDGEMENTS

The Institute thanks the authors of the papers, the editor, Dr. J.C.R. Hunt, AFIMA (University of Cambridge) and also Mrs. Janet Parsons, Mrs. Denise Bumpus, Miss Pamela Irving and Miss Karen Jenkins for typing the papers.

1. BASIC STUDIES AND MODELLING OF TURBULENCE AND DIFFUSION IN STRATIFIED FLOWS

DIFFUSION AND DECAY IN STABLY-STRATIFIED TURBULENT FLOWS

R.E. Britter

(Department of Engineering, University of Cambridge)

ABSTRACT

Laboratory experiments, by the author, on decaying grid turbulence and diffusion from a point source in stably-stratified turbulence are summarized. Further experiments are described in which a towed grid is used to determine the ratio of net potential energy change of a stably-stratified fluid to the work done by the grid on the fluid. This ratio, an integral form of a flux Richardson number, is found to be a decreasing function of the Froude number, $\dfrac{U}{NM}$.

INTRODUCTION

Laboratory experiments on decaying grid turbulence and diffusion from a point source in stably-stratified turbulence are described.

Much of the work presented at the IMA Meeting has since appeared as a paper in the Journal of Fluid Mechanics by Britter et al. (1983). In summary, the published work describes experiments in which a grid is towed horizontally along a large tank filled first with water and then with a stably-stratified saline solution. The decay rates of the r.m.s. velocity components (v', w') perpendicular to the mean motion are measured by a 'Taylor' diffusion probe and are found to be unaffected by the stable stratification over distances measured from 5 to 47 mesh lengths (M) downstream for Froude numbers $\dfrac{U}{NM}$ greater than 1.06, U being the velocity and N the buoyancy frequency. The Reynolds number $\dfrac{Mw'}{\nu}$ of the turbulence was about 10^3, where ν is the kinematic viscosity. The

vertical velocity fluctuations produced near the grid were
reduced by the stratification by up to 30% when $\frac{U}{MN} \simeq 0.5$.
Large scale internal wave motion was not evident from the
observations within about 50 mesh lengths of the grid.

The turbulent diffusion from a point source located 4.7 mesh
lengths downstream was studied. σ_y, σ_z, the horizontal and
vertical plume widths, were measured by a rake of probes. σ_y
was found to be largely unaffected by the stratification and
grew like $t^{\frac{1}{2}}$, while σ_z was found in all cases to reach an
asymptotic limit $\sigma_{z\infty}$ where $0.5 \lessgtr \frac{\sigma_{z\infty} N}{w'_s} \lessgtr 2$, w'_s being the r.m.s.
velocity fluctuations at the source; the time taken for σ_z to
reach its maximum was about $2N^{-1}$. The non-dimensional asymp-
totic plume depth $\frac{\sigma_{z\infty} N}{w'_s}$ was found to be proportional to $\left(\frac{U}{MN}\right)^{-\frac{1}{2}}$
or $\left(\frac{w'_s}{MN}\right)^{-\frac{1}{2}}$

The stable density-stratification of a fluid provides a
sink for turbulent kinetic energy in addition to viscous
dissipation. The partition of the energy supplied to these
two sinks has been of interest in several diverse contexts for
many years. Investigation has been hampered by the difficulty
of producing a "clean" experiment. One of the most recent
attempts, Linden (1980), in which a grid was dropped through
a density interface, was successful in providing a wide range
of the independent variable. However uncertainty entered in
deciding whether turbulence some distance from the interface
was really available for mixing the two fluids; the energy
sink resulting from the density stratification. A similar
experiment had been considered by this author using a linear
density-stratification. This was thought to be a "cleaner"
experiment until it was realised that all the profile changes
occur at the boundaries where the grid was either accelerating
or decelerating.

The present towed grid experiment provides a horizontal
version of the desired experiment.

Thus further work, not previously published, was described
at the IMA meeting in which the objective was to determine, in
a transient, towed grid experiment, the ratio of the potential
energy change in the density stratification in the towing tank
to the work done on the fluid by the passage of a grid through

the tank. This ratio was determined for a range of values of the dimensionless independent variable, the Froude number $\frac{U}{MN}$.

The force on the grid was not measurable so the principal uncertainty in this configuration is in assigning a drag coefficient to the bars of the grid.

EXPERIMENTAL PROCEDURE

The experiments were conducted in a towing tank having dimensions 1.2 m deep, 2.4 m wide, and 25 m long. The tank has an aluminum framework, and the sides and bottom are constructed with acrylic plastic for viewing purposes. Rails located on top of the side walls provide a support and track for the towing carriage. Models are attached to the carriage and can be towed the length of the tank at speeds in the range of 5 to 50 cm/s.

A salt water filling system was used to fill the tank to obtain various stable density profiles. The density profiles are determined by drawing water samples from various depths in the tank and measuring the specific gravity of each sample with a Mettler (PL 200) electronic balance. The specific gravity is measured by placing a weight, suspended from the electronic balance, into each water sample. Initially, the weight is placed in a sample of fresh water, and the output of the electronic balance is recorded. The specific gravity of each sample is then obtained by taking the ratio of the output for the sample and the output for fresh water. The tank was filled with salt solution to produce a linear density profile characterised by a Brunt Vaisala frequency, $N = [g(-\frac{\partial \bar{\rho}/\partial z}{\bar{\rho}})]^{-\frac{1}{2}}$ of 1.31 radians/sec.

A square-mesh, biplanar grid was constructed from rectangular, aluminum bars 1.9 cm wide and 1.3 cm thick with mesh size of 8.46 cm. Its solidity was thus 40%. The grid was rigidly attached to the towing carriage and towed the length of the tank in both directions with a constant speed.

The experiments were conducted by towing the grid over most of the length of the towing tank. A density profile was taken before initiating a series of tows. The grid was then towed back and forth through the tank at a constant speed always allowing an interval of time to pass between tows for the motion in the fluid to damp out until only slight gravity wave motion was visible. After a series of tows had been completed, the fluid was allowed to settle so that no motion was visible,

and a vertical density profile was obtained. The experiments were conducted over a range of four tow speeds; 6, 12, 30 and 49 cm/s.

The change in the vertical density profile, Fig. 1, shows that the original linear profile is maintained in the middle but eroded at the extremes. The symmetry of the experimental arrangement, away from the upper and lower edges, requires the maintenance of the linear profile. The profiles may be interpreted as the pumping of denser fluid up and less dense fluid down by the grid-generated turbulence.

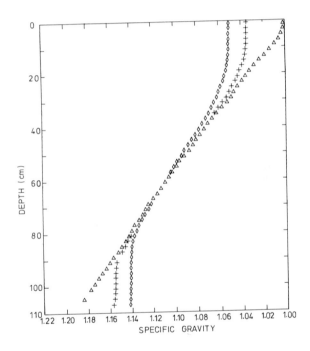

Fig. 1   Vertical profile of fluid specific gravity from before experiment, Δ and after + (U=30cm/s), ◇ (U=49cm/s).

PRESENTATION AND DISCUSSION OF RESULTS

The amount of work done on the fluid by towing the grid was calculated using the following equation:

$$W = \frac{S.G._{MD}\ \rho_{FW}\ C_D\ U^2\ A\ L_T}{2}$$

where $S.G._{MD}$ is the specific density of the water at mid-depth, ρ_{FW} is the density of fresh water, C_D is the drag coefficient for a rectangular bar (taken as equal to 2, see Turner, 1973, p. 152), A is the frontal-surface area of the grid (A = 11,020 sq. cm), and L_T is the total towing distance.

The potential energy change in the density stratification was calculated by doubling the change in potential energy in the density stratification over the top half of the water in the tank. It was felt that this would give a more representative value of the potential energy change since, for operational reasons, the grid did not extend right to the bottom of the tank.

Thus

$$\Delta P.E. = 2 g \rho_{FW} L W \int_0^{H/2} \Delta (S.G.) z \, dz$$

where the datum is taken as the mid depth in the tank and integration is over only the top half of the tank. L and W are the length and width of the tank respectively while H is the fluid depth. For later comparison purposes we note that W

equals $\bar{\rho} \int_0^\infty dt \iiint_V \varepsilon \, dV$ in the absence of density-

stratification, and that

$$\Delta P.E. = g \int_0^\infty dt \iiint_V \overline{w\rho} \, dV .$$

Also

$$W = \Delta P.E. + \rho \int_0^\infty dt \iiint \varepsilon \, dV$$

where the triple integral refers to the total fluid volume.

The ratio, T, of the potential energy change to the work input decreases, Fig. 2, from approximately 0.15 at $\frac{U}{MN} \simeq 0.5$ to 0.01 at $\frac{U}{MN} \simeq 5.0$. When the same data are replotted, Fig. 3, we note that they follow an inverse power law in this range.

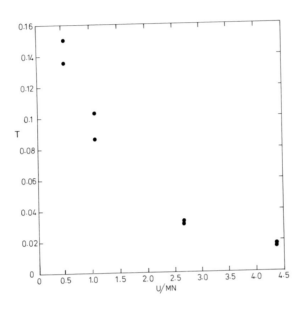

Fig. 2   The ratio T as a function of the Froude number $\dfrac{U}{MN}$

Using C_D = 1.6, as found experimentally by Linden, 1980, for a similar grid, the maximum observed value of T increased from 0.15 to 0.19.

The variable 1 - T represents the amount of turbulent kinetic energy dissipated by viscosity to heat and must always be finite. As the Froude number increases, that is when density-stratification becomes less important, a smaller proportion of the turbulent kinetic energy is responsible for increasing the potential energy of the system. Obviously T has an upper limit of unity, however, a smaller maximum might be expected by analogy with analysis and experiments for a steady, as distinct from a transient, problem, e.g. see Turner, 1973.

The ratio, T, is an integral form of a Richardson flux number which is typically defined as the ratio of the rate of potential energy change to the rate of production of turbulent kinetic energy. So T may be renamed the integral Richardson flux number Rf_I and plotted against the independent variable,

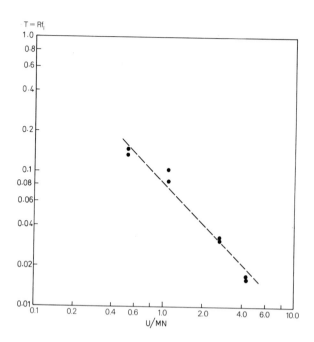

Fig. 3   The ratio $T = Rf_I$ as a function of the Froude

number $\frac{U}{MN}$.

the Richardson number $Ri_O = \frac{M^2 N^2}{U^2}$.   With these axes we see,

Fig. 4, that $Rf_I \simeq 0.085 \, Ri_O^{\frac{1}{2}}$.

In interpreting those results we might attempt to estimate
the flux resulting from the turbulence when the density-
stratification is very weak, that is $\frac{U}{MN}$ very large, $\frac{MN}{U}$ very
small.   In this limit the fluid may be treated as neutrally
stratified.   Thus the grid turbulence is described by

$$\rho \varepsilon = A_1 \, u^3 / \ell$$

where u and ℓ are the turbulent velocity and length scales and
A_1 is an empirical constant of order unity.

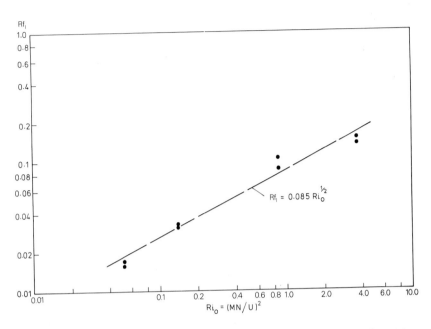

Fig. 4   The integral Richardson flux number Rf_I as a function
of the Richardson number, $Ri_O = (\frac{MN}{U})^2$.

The density flux may be estimated with

$$\overline{w\rho} = - K \frac{\partial \bar{\rho}}{\partial z}$$

where the eddy diffusivity K is equal to A_2 u ℓ.   Thus

$$\overline{w\rho} = A_2 \, \bar{\rho} \, \frac{u \, \ell \, N^2}{g}$$

and the ratio

$$\frac{g\overline{w\rho}}{\rho\varepsilon} = \frac{A_2}{A_1} \cdot \frac{N^2\ell^2}{u^2} = A_3 \, \frac{N^2M^2}{U^2} \left(\frac{\ell^2}{M^2}\right)\left(\frac{u^2}{U^2}\right)^{-1}$$

which will scale on $\dfrac{N^2 M^2}{U^2}$. For a transient experiment the analogous ratio is

$$\frac{\displaystyle\int_{O}^{\infty} \overline{gw\rho}\ dt}{\displaystyle\int_{O}^{\infty} \overline{\rho\varepsilon}\ dt}$$

However the numerator is unbounded and Rf_I, as defined, is an apparently inappropriate variable in the passive limit. Much the same difficulty was encountered in Linden's (1980) experiments and arguments based on negligible diffusion after the "final" period of decay were required. Recourse to similar

arguments here lead to $Rf_I \ \alpha \ \left| Ri_O = \dfrac{M^2 N^2}{U^2} \right|$ but with the coeffi-

cient subject to, somewhat, arbitrary specification.

There is currently considerable interest in how a Rf-Ri curve behaves near its maximum. The question concerns whether the maximum is maintained - a critical Richardson flux number condition or whether Rf decreases with increasing Ri (Linden, 1980). Unfortunately the present data provide no information on this problem. Larger values of Ri_O are not attainable together with maintenance of a flow homogeneity. The wakes of the horizontal bars will never interact with each other at higher Ri_O.

FURTHER DISCUSSION

Pearson et al. (1983) use a Lagrangian model to derive the density flux as

$$F_\rho = \frac{1}{2}\ \frac{d<z^2>}{dt}\ \left(-\frac{\partial\overline{\rho}}{\partial z}\right) + <\Delta\rho(t)w(t)>$$

where the < > brackets refer to fluid particles. The first term represents fluid particle displacement while the second term represents molecular diffusion of species between the particle and the surrounding fluid. They then consider various models of these terms.

A simpler, speculative approach, applied to the present transient experiment, is to view the mixing process as (a) fluid particles being removed from their equilibrium position in the absence of molecular diffusion and then (b) being trapped there either by (i) molecular diffusion of species or (ii) decay of the turbulence.

Although the effects described will occur simultaneously a simple sequential model seems worthy of investigation.

Thus we initially neglect molecular diffusion and approximate F_ρ with

$$\frac{1}{2} \frac{d<z^2>}{dt} \left(-\frac{\partial \bar{\rho}}{\partial z} \right)$$

so that

$$\int_0^\infty F_\rho \, dt = \frac{1}{2} <z_\infty^2> \left(-\frac{\partial \bar{\rho}}{\partial z} \right).$$

Substitution into the definition of Rf_I produces

$$Rf_I = \frac{g<z_\infty^2> \left(-\frac{\partial \bar{\rho}}{\partial z} \right) L \, WH/2}{\rho \, C_D U^2 AL_T/2}$$

$$= \frac{<z_\infty^2> N^2}{U^2} \left[\frac{L}{C_D L_T} \cdot \frac{WH}{A} \right]$$

where the final bracketed term is a constant.

Britter et al. (1983) showed that $\dfrac{\sigma_{z\infty} N}{w'_s}$ was proportional to $\left(\dfrac{U}{MN}\right)^{-\frac{1}{2}}$ where $\sigma_{z\infty}$ is the maximum plume width resulting from a point source downstream of the grid. Although this will be less than $<z_\infty^2>^{\frac{1}{2}}$, which refers to particle displacements from an imaginary source at the grid, they should be proportional to each other. Attaching a typical fluctuating velocity to the flow at the grid is difficult. However, provided the flow at the grid plane is not influenced by the density-stratification (that is the Froude numbers based on the bar width is large) the fluctuating velocities leading to diffusion will scale on U. Thus $\dfrac{<z_\infty^2>^{\frac{1}{2}} N}{U}$ for marked particles released at the grid

will be proportional to $(\frac{U}{MN})^{-\frac{1}{2}}$. Therefore Rf_I is expected to be proportional to $(\frac{U}{MN})^{-1}$, or $Ri_0^{\frac{1}{2}}$, consistent with our observations.

This apparent link between suitably defined Richardson flux numbers and turbulent diffusion from a point source is, with hindsight, not so surprising. Further, Pearson et al. (1983) present calculations that require an upper limit to $\frac{\sigma_{z\infty} N}{w'_s}$ as $\frac{U}{MN}$ is reduced. This upper limit should be associated with an upper limit to Rf_I, a critical Richardson flux number, but further study is required.

REFERENCES

Britter, R.E., Hunt, J.C.R., Marsh, G.L. and Snyder, W.H. (1983) The effects of stable stratification on turbulent diffusion and the decay of grid turbulence. *Jnl. Fluid Mech.*, **127**, pp. 27-44.

Linden, P.F. (1980) Mixing across a density interface produced by grid turbulence. *Jnl. Fluid Mech.*, **100**, pp. 691-703.

Pearson, H.J., Puttock, J.S. and Hunt, J.C.R. (1983) A statistical model of fluid element motions and vertical diffusion in a homogeneous stratified flow. *Jnl. Fluid Mech.*, **129**, pp. 219-249.

Turner, J.S. (1973) Buoyancy effects in fluids. Cambridge University Press.

THE INHIBITION OF VERTICAL TURBULENT SCALE
BY STABLE STRATIFICATION

F.K. Browand

(University of Southern California)

and

E.J. Hopfinger

(Université de Grenoble)

INTRODUCTION

In the past few years, laboratory studies and field observations have given indications that several relatively simple ideas may be usefully applied to stratified turbulence. The first idea - most clearly expressed in the oceanic context by Osmidov (1965) - is that stratification will act to limit the largest scales in the flow, i.e., there is some maximum eddy size for the turbulence. An estimate for this "Osmidov Scale", or buoyancy scale, can be made if the energy, scale of the turbulence, and the stratification are known. If the buoyancy scale is much larger than the turbulence integral scale, then stratification will have a relatively minor effect on the turbulence. In many situations, however, the turbulent energy - and thus the buoyancy scale - decreases with time, while the integral scale of the turbulence increases. A point will be reached for which the buoyancy scale and turbulence scale are comparable. This coincides with the onset of the influence of stratification. The onset is rapid and dramatic - severely limiting the largest scales and therefore the turbulent entrainment. The shear flow experiment of Koop and Browand (1980) illustrates the rapidity of the onset of stratification effect. Lange (1974) also suggested a rapid alteration of turbulence structure (when integral scale becomes comparable to buoyancy scale) on the basis of his towed grid experiments. The wake collapse experiments described by Lin and Pao (1979), and the experiments of Dickey and Mellor (1980) and of Stillinger (1981) are in agreement on this point.

Osmidov further suggested that a decaying patch of turbulence in the ocean will support a range of active scales between the buoyancy scale and the Kolmogoroff scale (see also Gibson

(1980)). As time progresses, the largest scale decreases and the smallest scale increases - hence the range of active turbulent scales becomes narrower and narrower. When the buoyancy scale reaches the Kolmogoroff scale, turbulence ceases and the flow must consist of internal waves and slow, lateral, intrusive features. Dillon and Caldwell (1980) have measured the smallest scales of temperature fluctuation in the oceanic mixed layer. Within the range of smallest scales, they conclude that stratification affects the largest scales first. That is, departures from the Batchelor spectrum occur first at lowest wave number. This is consistent with the preceding view.

The buoyancy scale can be obtained by dimensional reasoning when turbulent down-scale energy flux (equal to dissipation rate), ε, and stratification frequency, N, are the only important parameters characterizing the turbulence. One obtains

$$\ell_B \cong (\varepsilon/N^3)^{\frac{1}{2}} \cong \left(\frac{u'^3/\ell}{N^3}\right)^{\frac{1}{2}} \tag{1}$$

where u' is the turbulent intensity, and ℓ is the longitudinal integral scale. (In homogeneous, isotropic turbulence, ε and u'^3/ℓ are related by a constant of order unity.) A physical argument is based on the requirement that a turbulent eddy must have sufficient momentum to execute an eddy-like motion in the vertical plane against the torque supplied by the buoyancy force, as heavier elements are lifted and lighter elements are depressed (Fig. 1). The largest scale for which sufficient momentum exists is

$$\ell_B = \frac{u'}{N}$$

which is consistent with (1) upon setting $\ell = \ell_B$.

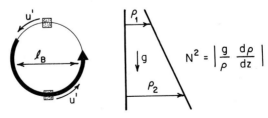

Fig. 1   Sketch illustrating largest turbulent eddy scale which has sufficient momentum to overturn against the buoyancy torque

A simple experimental verification of maximum vertical scale
was devised in Grenoble in the fall of 1980. The experiment
consisted of a square-bar grid resting vertically in a tank
containing salt stratified fluid with a linear density
variation. At time t = 0, the grid was set into oscillation at
frequency f, producing a vigorous, turbulent region. Initially
the turbulent integral scale was smaller than the maximum
allowable scale. As the front spread laterally, the buoyancy
scale decreased and the integral scale increased. At some
point, the two must cross, as depicted in Fig. 2, and visual
observations of frontal shape should reflect the increased
importance of buoyancy.

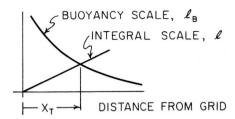

Fig. 2   Sketch of buoyancy scale and integral scale variation
         with distance away from grid. Distance, x_T, is point
         where scales have comparable magnitude

Recently, Ivey and Corcos (1982) and Thorpe (1982) have
published experiments using similar geometries and techniques.
The results of all three experiments appear to be substantially
consistent, although we have adopted a slightly different point
of view in our interpretation.

THEORETICAL PREDICTIONS

One expects to observe changes in frontal shape at the
position where the integral scale and the buoyancy scale are
comparable. This point is termed x_T in Fig. 2. A quantitative
expression for x_T can be obtained by utilizing the results of
Hopfinger and Toly (1976) for steady, grid generated turbulence
in unstratified fluid. The use of these results involve two
assumptions. The flow in the initially turbulent region
$x < x_T$ must not be appreciably affected by buoyancy; and the
local turbulence quantities behind the advancing front must
behave as they would in the steady flow at the same value of x.
It has already been remarked that the onset of stratification
effect is rapid, so the first assumption seems   reasonable.
There is little information to support or refute the second

assumption, although Long (1978) and Dickinson and Long (1978) argue that the spread of a turbulent front into unstratified fluid (proportional to $(t)^{\frac{1}{2}}$), implies constancy of the product $u'\ell$ - as observed in the steady problem.

The results of Hopfinger and Toly are:

$$u' \cong \frac{(.29)S(SM)^{\frac{1}{2}}f}{x} = \frac{(.046)S(SM)^{\frac{1}{2}}\omega}{x} \tag{2}$$

$$\ell \cong (.26)x \tag{3}$$

with x measured from the midplane of the oscillating grid. The constants in (2) and (3) undoubtedly depend upon the particular grid geometry. For these experiments, the single-plane, square-bar grid had mesh M = 5 cm, bar size d = 1 cm, and stroke S = 4 cm. Then let

$$\ell_B = \left(\frac{u'^{3}/\ell}{N^3}\right)^{\frac{1}{2}}$$

define the buoyancy scale, and require that

$$\ell_B = A\ell_T \quad \text{at } x = x_T$$

where ℓ_T is the value of the integral scale at $x = x_T$, and A is a multiplicative constant. The resulting predictions are:

$$x_T = \frac{(1.80)}{A^{\frac{1}{3}}} \left(\frac{\omega}{N}\right)^{\frac{1}{2}} \tag{4}$$

$$\ell_T = \frac{(.47)}{A^{\frac{1}{3}}} \left(\frac{\omega}{N}\right)^{\frac{1}{2}} \tag{5}$$

with x_T, ℓ_T expressed in centimetres.

EXPERIMENTAL VERIFICATION

The experiments were performed in an enclosed tank measuring 50 cm deep, 30 cm wide, and 200 cm in length (Fig. 3). The grid was positioned at the middle of the tank. Grid mesh size, bar size, and stroke were respectively, 5 cm, 1 cm, and 4 cm - identical to the earlier experiments of Hopfinger and Toly. Two grids were used - one having grid bars oriented along 45 degree diagonals with respect to horizontal the other having horizontal and vertical bars. The results are the same for the two grids. The walls of the tank were treated as image planes with respect to the grid geometry - although this was more accurately done for the grid consisting of horizontal and vertical bars. A gap of approximately 1 mm existed between the ends of the bars and the walls of the tank. Grid oscillation frequency varied between 9.0-35. rad/sec, and values of N were in the range O, and 0.35-1.0 rad/sec. Reynolds numbers for the turbulence near the grid - based on mesh size - were 8000 - 20 000.

A simple shadowgraph image was produced on translucent paper attached to the front tank wall (Fig. 3). In addition, potassium permanganate dissolved in salt water was used as a dye and added to the fluid near the grid just prior to the start of oscillation. This combination gave the best defini- tion of the turbulent fluid front. The front is not uniform across the width of the tank, and no attempt was made to characterize the horizontal variations. The image we observed is an average which emphasizes the more coherent features. A sequence of turbulent front positions were recorded on film at times measured from the beginning of oscillation. From these negatives, turbulent front positions were first traced and later digitized on a large, Hewlett Packard digitizing table.

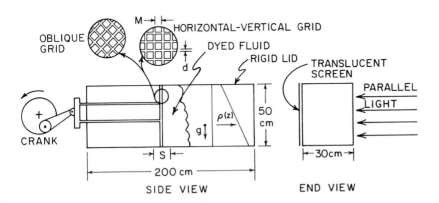

Fig. 3   Experimental apparatus

Several photographs of the interface are shown in Fig. 4, taken at non-dimensional times Nt = 2.7, 29.8, for ω/N = 15.4. It is typical of the results obtained. Fig. 5 gives sequences of digitized front positions for three experiments corresponding to different initial values of the ratio ℓ_B/ℓ. The first experiment, 5(a), has no stratification and corresponds to $\ell_B/\ell = \infty$. Interface distortions (or bulges) evolve rapidly to form scales comparable to the depth of the tank. With stratification 5(b), (c), the scale of interface bulges does not continue to grow. The interface changes character, evolving into a series of intrusions which continue to lengthen as time progresses, but do not increase in vertical extent. The more energetic the initial turbulence - as measured by the initial value of ℓ_B/ℓ - the larger the scale of the intrusive features.

It is possible to visually identify x_T, the position of the interface when the intrusions first become evident. A more objective criterion for the formation of intrusions is obtained by computing a non-dimensional measure of the shape of the interface. The simplest shape parameter is the ratio of the variance of the front irregularity, divided by the mean front position, averaged over the length of the interface (depth of the tank). This ratio often (but not always) increases sharply as the front begins to differentiate into layers. This point is taken to mark the onset of stratification effect. Estimates of the transition distance, x_T, are plotted as a function of ω/N in Fig. 6. Both visual estimates and estimates based upon the shape parameter are plotted to give some idea of reliability. The results are not substantially different.

The dependence upon ω/N can be fit by a line of slope 1/2, in agreement with the simple theory based upon a coincidence of the integral scale and the buoyancy scale. The experimental result is approximated by

$$x_T \cong (3.5) \left[\frac{\omega}{N}\right]^{\frac{1}{2}}$$

which can be used to evaluate the constant A in (4), A = .136.

One would also expect the vertical scale of the intrusions to be proportional to the integral scale, ℓ_T, at the point $x = x_T$. This is borne out by plotting the vertical thickness of intrusions as a function of ω/N, in Fig. 7. The vertical

Fig. 4 Interface development for $\omega/N = 15.4$; (a) $Nt = 2.7$;
(b) $Nt = 29.8$

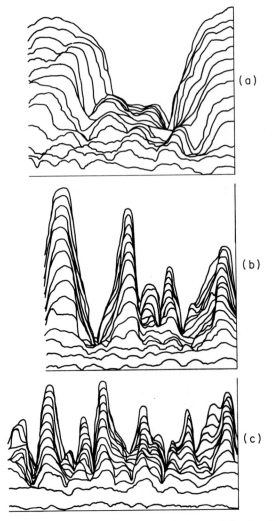

Fig. 5 Sequence of digitized front positions (a) $(\ell_B/\ell)_i = \infty$;
(b) $(\ell_B/\ell)_i = 31.0$; (c) $(\ell_B/\ell)_i = 19.0$

dimension, ℓ_v, is determined at the mean position of the interface, $x = x_T$, for all the intrusions which exceed a threshold
length of approximately 0.5 times the root-mean-square displacement of the interface. The threshold was used to ensure
that only well-formed intrusive features were selected by the
computer. The mean intrusion scale can be fit by

$$\bar{\ell}_v \cong (1.0) \left(\frac{\omega}{N}\right)^{\frac{1}{2}}$$

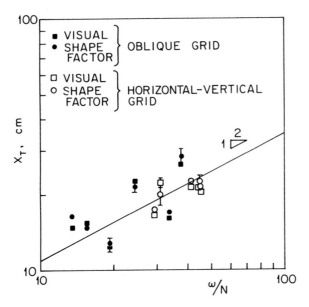

Fig. 6   Experimental values of transition distance, x_T, versus ω/N

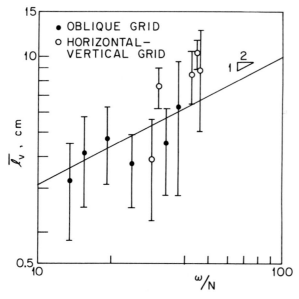

Fig. 7   Experimental values of mean transition thickness at $x = x_T$. Bars denote the estimated variance

which, from (5), implies that $\bar{\ell}_v$ is a multiple of ℓ_T.

$$\ell_v \cong (1.1)\,\ell_T$$

DISCUSSION

The conclusion to be drawn from the data presented is that the buoyancy scale, ℓ_B, does indeed limit the possible vertical scales in the turbulent flow. For this particular geometry, the relationship between the three scales ℓ_B, ℓ_T, $\bar{\ell}_v$, at the point of arrest, are

$$\ell_T \cong 7.5\,\ell_B$$

$$\bar{\ell}_v \cong 8.\ell_B$$

There are several reservations, however. Referring again to Figs. 6 and 7, one sees that the scatter in the results is considerable. The choice of the method in determining x_T does not account for the scatter. No explanation is obvious - it may simply be a consequence of the turbulent nature of the flow. More data over a larger range of (ω/N) would give greater confidence in the result. Second, the value of the constant, A, is smaller than one might expect on physical grounds. One of the experimental short-comings is that local turbulence quantities were not measured, but inferred from related experiments. Also, the definition of the transition point is somewhat arbitrary. All these uncertainties accumulate in the constant A. We feel the present results (and also those of Thorpe) support the general conclusion, but the actual value of the constant A should be treated with caution.

The manner of spread of the intrusive features is also of considerable interest. The intrusions do not have uniform density - that is, they are not well-mixed, but only partially mixed. The actual density gradient measured across the thick-ness of a layer is in part a result of the irreversible (molecular) mixing which has taken place near the grid, and in part a result of the motion of the intrusion itself. The intrusion advances, causing induced motion and induced density structure. (If the flow is altered by stopping the grid, a portion of the density anomaly relaxes).

Fig. 8 Propagation of soliton-like waves on individual
        intrusion, ω/N = 19.25, N = .643 rad/sec. Time interval
        between each photo is 2 seconds

A series of solitary wave-like disturbances are observed to progress along the cores toward the tips of the individual intrusions. The waves seem to be generated by the turbulence near the point $x = x_T$. Fig. 8 shows a sequence of photographs of the progression of a mode 2 wave propagating left-to-right on one of a number of similar intrusions. When the waves arrive at the tip, the induced motion causes the front to advance.

There is an analogy between this flow constrained by stratification and the rotating flow studied by Hopfinger, Browand and Gagne (1982).

Both flows have turbulent scales limited by the application of a body force (either Coriolis force or buoyancy force). In both cases, one observes non-linear waves generated by the turbulence. These waves seem to be important in providing the mechanism of energy transport between the strongly turbulent region near the grid and the less turbulent region farther away.

ACKNOWLEDGEMENTS

The authors gratefully express thanks for support from the Office of Naval Research, Fluid Mechanics and Oceanography Programs; and from the C.N.R.S., Program of Oceanography.

REFERENCES

Dickey, T.D. and Mellor, G.L. (1980) Decaying Turbulence in Neutral and Stratified Fluids, *J. Fluid Mech.*, **99**, 13-32.

Dickinson, S.C. and Long, R.R. (1978) Laboratory Study of the Growth of a Turbulent Layer of Fluid, *Phys. Fluids*, **21**, 1698-1701.

Dillon, T.M. and Caldwell, D.R. (1980) The Batchelor Spectrum and Dissipation in the Upper Ocean, *J. Geophys. Res.*, **85**, 1910-1916.

Gibson, C.H. (1980) "Fossil Temperature, Salinity, and Vorticity in the Ocean", Marine Turbulence, (J.C.T. Nihoul, ed.), Elsevier.

Hopfinger, E.J. and Toly, J.A (1976) Spatially Decaying Turbulence and its Relation to Mixing Across Density Interfaces, *J. Fluid. Mech.*, **78**, 155-176.

Hopfinger, E.J., Browand, F.K. and Gagne, Y. (1982) Turbulence and Waves in a Rotating Tank, *J. Fluid. Mech.*, **125**, 505-534.

Ivey, G.N. and Corcos, G.M. (1982) Boundary Mixing in a Stratified Fluid, *J. Fluid Mech.*, **121**, 1-26.

Koop, C.G. and Browand, F.K. (1979) Instability and Turbulence in a Stratified Fluid with Shear, *J. Fluid Mech.*, **93**, 135-159.

Lange, R.E. (1974) Decay of Turbulence in Stratified Salt Water, Ph.D. Thesis, University of California, San Diego.

Lin, J.T. and Pao, Y.H. (1979) Wakes in Stratified Fluids, *Ann. Rev. of Fluid Mech.*, **11**.

Long, R.R. (1978) Theory of Turbulence in a Homogeneous Fluid Induced by an Oscillating Grid, *Phys. Fluids,* **21**, 1887-1888.

Ozmidov, R.V. (1965) On the Turbulent Exchange in a Stably Stratified Ocean, *Izv. Atmos. Ocean. Phys.*, **1**, 853-860.

Stillinger, D.C. (1981) An Experimental Study of the Transition of Grid Turbulence to Internal Waves in a Salt-Stratified Water Channel, Ph.D. Thesis, University of California, San Diego.

Thorpe, S.A. (1982) On the Layers Produced by Rapidly Oscillating a Vertical Grid in a Uniformly Stratified Fluid, *J. Fluid Mech.*, **124**, 391-409.

TURBULENCE AND WAVE MOTIONS NEAR AN INTERFACE BETWEEN A TURBULENT REGION AND A STABLY STRATIFIED LAYER

D.J. Carruthers and J.C.R. Hunt

(DAMTP, University of Cambridge)

SUMMARY

Linear theory is used to study velocity fluctuations near an interface between a turbulent region and a stably stratified layer, when the energy dissipation rate is approximately constant with height in the turbulent region, and when there is no mean shear.

Variances of the fluctuating velocities, one dimensional spectra and integral scales are calculated in both regions, and the wave flux is calculated in the stratified layer. A prediction of the theory is that eddies with frequency of order of the buoyancy frequency (N) of the stratified layer are least affected by the stratification. In the stratified layer, since waves with frequency $\omega > N$ decay rapidly with the distance z from the interface, the high frequency parts of the spectra fall off sharply, a striking feature of the atmospheric measurements of Caughey and Palmer (1979).

In the limit of large stratification ($N \to \infty$), the theory shows that the effect of the stable layer on the turbulent region is the same as that of a rigid surface moving with the flow at the same mean velocity (i.e. the solution of Hunt and Graham, 1978). In the limit of small stratification ($N \to O$), the vertical motions in the turbulent region decrease in intensity near the interface and irrotational motions are induced in the slightly stable layer (i.e. the same result as Philips', 1955). Some preliminary results are presented for the common situations where there is a thin layer of strong stable stratification at the interface between the stable layer and the turbulent region. We also discuss how the analysis compares with other computations and turbulence models.

1. INTRODUCTION

There are many natural flows where regions of turbulent motion adjoin regions of stably stratified fluid in which there is no local production of turbulence. These include the upper part of the atmospheric convective boundary layer, which is capped by an inversion, the oceanic mixed layer which is bounded by the thermocline, and the 'benthic' boundary layer on the ocean bottom (Richards q.v.). These flows also occur where layers of turbulent flows in the environment are stably stratified artificially, such as by the discharge of heated water (Rogers and Pickles q.v.) or dense gases (Hunt et al., 1984).

The interaction between the turbulent region and stratified layers needs to be better understood because it affects and often controls the movement of the interface (dz_i/dt) and fluxes of momentum (F) and scalar quantities (F_s) across the interface. For instance, in large-scale numerical model studies of the atmosphere, these quantities are estimated from physical arguments and experimental data in terms of various local physical quantities. It is particularly important to estimate fluxes of heat and water vapour across those interfaces in determining the development of stratocumulus clouds, which are a common feature of the upper part of the convective boundary layer (Caughey et al., 1982).

It is convenient to subdivide the general class of turbulent flows confined by stratification into three main types of flow all of which occur frequently (Fig. 1). We assume there is no potential temperature gradient in the turbulent layer.

(a) The stable layer is uniformly stratified as in the laboratory experiments of Willis and Deardorf (1974), and the atmospheric measurements of Caughey and Palmer (1979). There is no mean shear.

(b) There is an intensely stratified layer marking the edge of the turbulent layer with fluid of lower stratification (which may be neutral) beyond e.g. atmospheric measurements of Caughey et al. (1982). There is no mean shear.

(c) Either of the above cases with mean shear. This shear is usually most intense when it is associated with a narrow stably-stratified layer (Brost et al., 1982).

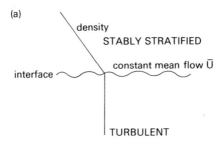

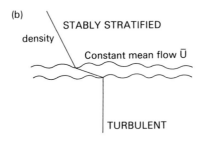

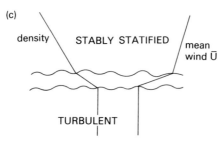

Fig. 1 Pictorial representation of the flow fields in which
turbulence is confined by stratified layers.

(a) Single stratified layer; no wind shear.

(b) A strongly stratified layer bounds the turbulence
with less stratified fluid beyond; no wind shear.

(c) A mean shear exists across the strongly stratified
layer.

A large number of laboratory experiments have been performed to measure the rate of growth of the thickness (z_i) of the turbulent layer and the fluxes for each of the above cases. Both these kinds of measurements are often expressed as 'entrainment velocities' E, where $E = dz_i/dt$ or $E = F_s/\Delta S$, where ΔS is the change in the value across the layer. Numerous attempts have been made to relate these entrainment velocities to the bulk properties of the flow (e.g. Kato and Phillips, 1969, Kantha, Phillips and Azad, 1978, Price, 1979, Turner, 1973, Willis and Deardorff, 1974, Piat and Hopfinger, 1981).

Four main mechanisms have been proposed for the entrainment process (Fig. 2).

(i) Turbulent eddies impinge on the interface and generate sufficiently large fluctuating velocity gradients that the local Richardson number $g \dfrac{\partial\theta/\partial z}{\theta(\partial u/\partial z)^2}$ is small enough for Kelvin-Helmholtz billows to grow and break and induce molecular mixing. One expects this process to be relevant when stratification is strong enough to damp the vertical component of turbulence at the interface; several authors (e.g. Long, 1978) have commented that in this limit fluctuations near the interface are expected to be similar to those in a turbulent flow near a rigid surface in zero mean, shear as described by the theory of Hunt and Graham (1978).

(ii) With strong stratification and energetic turbulence, eddies impinging on to the interface distort it sufficiently that fine filaments of the stratified fluid layer are drawn into the turbulent region where again molecular diffusion completes the mixing process (Linden, 1973).

(iii) Turbulent eddies distort the interface and set up internal waves in the stratified layer whose energy and form depend on the stratification. For a uniformly stratified layer (Fig. 1), the waves propagate energy away from the interface but the wave amplitude at the interface may be large enough to induce mixing. If the stratified layer is strong near the interface and weak or non-existent far above the interface, then trapped and resonant waves of large amplitude can be induced by quite weak turbulence.

(iv) With weak stratification the turbulent layer can also grow and entrain by the same processes as occur at the edge of a turbulent boundary layer or wake in neutral stratification; the large eddies in the turbulent layer induce large random

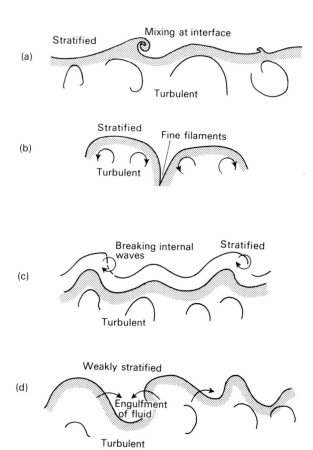

Fig. 2 Mechanisms of entrainment

(a) Mixing due to fluctuating velocites at an interface.

(b) Fine filaments entrained by energetic eddies.

(c) Waves breaking in stratified fluid.

(d) Engulfment of fluid of weak stratification.

motions in the external layer leading to the engulfment of
external fluid. In the limit of zero stratification, the
velocity fluctuations outside the turbulent interface are
irrotational (Phillips (1955); Kovasnay, et al. (1970)).

In this paper we do not aim to provide a new model for
entrainment, but we present the results of a linear analysis
for how the turbulence distorts the interface and induces wave
motions and irrotational motions in the stratified layer, and
also for how the stratified layer distorts the turbulence.
Previously Townsend (1966) and Stull (1976) calculated the
wave motions above a convective layer, but they simply assumed
the form of the vertical velocity fluctuations at the
interface. They did not attempt to match the convective
turbulence to the wave motion. We shall concentrate on case
(a) above in which the turbulence is confined by a single
uniformly stratified layer and there is no mean shear. However
the analysis has also been extended to case (b) in which there
is a shallow but strong inversion bounding the turbulence with
a less stable layer beyond, and we shall refer to preliminary
results for this case. The method of analysis is described
in detail in Carruthers and Hunt (1984) subsequently referred
to as CH. In this paper we emphasize the physical assumptions
required and the physical insights gained from the analysis.

We also compare calculations with observations of velocity
and temperature (or density) fluctuations. Linear analysis
may be appropriate for such a comparison, because much of the
time these fluctuations are associated with small deformations
of the interface. But the linear analysis is not appropriate
for calculating the nonlinear mixing process involved in
entrainment because these are often associated with large
deformations of the interface. However, the theory may provide
a method for identifying under what circumstances which of the
above mechanisms may control entrainment, since all the
mechanisms are sensitive to the local fluctuating velocity
field near the interface.

2.   THEORY AND RESULTS FOR CASE (a)

2.1 Assumptions and equations

Fig. 1(a) is a schematic representation of the structure
for which we shall outline the theory.   In the atmosphere
layer 1 represents the convective boundary layer whilst layer
2 represents the capping stratified layer.

(i)   Turbulent region

   In the interior of the layer the turbulence is assumed
to be homogeneous and to be specified by its longitudinal
integral length scale L_H, a typical r.m.s. velocity scale U_H
and the form of the three dimensional wavenumber spectrum
$\phi_{ij}^{(H)}(\underline{K})$.   Following the observations in the atmospheric
boundary layer by Caughey and Palmer (1979) and Kaimal et al.
(1976) we assume that the energy dissipation rate (ϵ) is
approximately constant with height.   It can then be shown for
high turbulent Reynolds numbers (Hunt, 1984) that

$$\frac{\partial \bar{\omega}^2}{\partial z} = 0 \qquad (2.1)$$

where $\underline{\omega} = \underline{\nabla} \times \underline{u}$ is the   vorticity, and the distortion of the
small scales of turbulence is irrotational.   Then assuming
incompressibility the simplest solution for the turbulence
field   satisfying (2.1) is

$$\underline{u} = \underline{u}^{(H)} - \nabla\phi, \qquad (2.2)$$

where $\underline{u}^{(H)}$ is a homogeneous turbulence field and $\nabla^2\phi = 0$, so
that $\nabla\phi = \underline{u}^{(s)}$ is an irrotational velocity field.   Note that
superscript (H) is used to denote a variable function in the
homogeneous region whereas subscript H is used to denote a
constant.

(ii)   Stratified region

   In this region the fluid is uniformly stratified with
buoyancy frequency N.   We solve the linearized incompressible
Boussinesq equations (CH).   As $z \to \infty$ the radiation condition is
applied for the intrinsic frequency of the wave motion $\omega < N$,
whilst for $\omega > N$ then the vertical velocity fluctuation $w \to 0$.

(iii)   At the interface

   The solutions in the two regions must be matched at the
interface ($z=0$).   The Kinematic Condition is that w is
continuous on $z=0$ whilst assuming continuity of pressure it
follows that $\frac{\partial w}{\partial z}$ is also continuous.   In order to solve the
system of equations, the frequencies with which waves in the
stratified region are excited by the different scales of

turbulence must be stipulated (CH). Hence we require an estimate of the joint wavenumber and frequency spectrum of the homogeneous turbulence $\chi_{ij}^{(H)}$ $(\underline{K},\omega)$ in terms of the wavenumber spectrum of the homogeneous turbulence $\phi_{ij}^{(H)}$ $(\underline{K})$, which is assumed to be known. This is achieved by using a hypothesis of Tennekes (1975) that the random advection of eddies by the most energetic eddies (on a time scale $\dfrac{l}{u_H}$ where l is the length scale of the advected eddy) results in an 'Eulerian' frequency spectrum with greater amplitude than the 'Lagrangian' frequency spectrum resulting from the change of the velocity of a fluid element passing a point moving with the mean flow (which occurs in the Lagrangian time scale). It can then be shown (CH) that the wavenumber frequency spectrum has the approximate form

$$\chi_{ij}^{(H)}\ (\underline{K},\omega)\ =\ \phi_{ij}^{(H)}\ (\underline{K})\,\delta\,(\omega-u_H k) \qquad (2.3)$$

where $k = |\underline{K}|$. In the analysis it was found convenient to use a modified form of (2.3)

$$\chi_{ij}^{(H)}\ (\underline{K},\omega)\ =\ \phi_{ij}^{(H)}\ (\underline{K})\,\delta\,(\omega-u_H k_{12}) \qquad (2.4)$$

where $k_{12} = (K_1^2 + K_2^2)^{\frac{1}{2}}$, but this does not change the form of the results (CH).

(iv)   Method of solution

The equations and matching conditions at the interface can be used to calculate the one dimensional spectra, variances and integral scales in both the turbulent and stratified layers in terms of spectra, variances and scales of the homogeneous turbulence. The method is described in detail by (CH); we shall not repeat this analysis here but present the most interesting results.

2.2 General results for any spectra of the homogeneous turbulence

A number of results can be obtained without specifying the turbulence spectrum. There is a discontinuity in the one dimensional spectra, $\Theta_{jj}(K_1)$, of the velocity components at the interface given by

$$\Theta_{jj}(K_1, z=0^-) = \Theta_{jj}(K_1, z=0^+) + \Theta_{jj}^{(H)}(K_1) \qquad (2.5)$$

where $j = 1$ or 2; this is independent of the buoyancy frequency N. At low horizontal wavenumbers since $\underline{u}^{(s)} = -\nabla\phi$, $\hat{u}^{(s)}(K_1=0) = \hat{v}^{(s)}(K_2=0)$. Thence

$$\Theta_{11}(K_1=0, z=0^-) = \Theta_{11}^{(H)}(K_1=0) \text{ and } \Theta_{22}(K_2=0, z=0^-) = \Theta_{22}^{(H)}(K_2=0) \qquad (2.6)$$

and

$$\Theta_{11}(K_1=0, z=0^+) = \Theta_{22}(K_2=0, z=0^+) = 0 \qquad (2.7)$$

Thus the low wavenumber ends of the longitudinal spectra are independent of N and z.

In the stratified region the gravity wave parts of the spectra ($\omega < N$) are independent of z; when $N \to \infty$

$$\Theta_{33}(K_1) = 0 \qquad (2.8)$$

and

$$\overline{u_1^2} = \overline{u_2^2} = 0.5\, \overline{u_H^2} \qquad (2.9)$$

In the limit $N \to \infty$ solutions for $\Theta_{jj}(K_1, j=1,2,3)$ in the turbulent region are identical to those of Hunt and Graham (1978) for turbulence near a rigid surface, thus in this limit the turbulence below the interface behaves as if it were confined by a rigid surface.

In the limit of zero stratification $N \to 0$, the solution in
the non-turbulent region ($z \geqslant 0^-$) is similar to Phillips'
(1955) solution for irrotational fluctuations outside a turbulent region. However Phillips did not consider any feedback
between the two layers. He used only the kinematic boundary
condition at the interface and assumed that the turbulence
was undistorted by the fluid in the irrotational region,
whereas the absence of rotational motion above $z=0$ must reduce
the velocity fluctuations below $z=0$. The difference between
these solutions only affects the magnitude of the irrotational
fluctuations in terms of the magnitude of the homogeneous
turbulence; it does not affect the relative magnitudes of the
different components of the irrotational velocity fluctuations.
Our solution gives

$$\overline{u}_3^2(z=0) = 0.25 \; u_H^2$$

$$\hspace{4cm} (2.10)$$

$$\theta_{33}(K_1, z=0) = 0.25 \; \theta_{33}^{(H)}(K_1),$$

whereas in Phillips' (1955) solution

$$\overline{u}_3^2(z=0) = u_H^2$$

$$\hspace{4cm} (2.11)$$

$$\theta_{33}(K_1, z=0) = \theta_{33}^{(H)}(K_1),$$

and both results are independent of the form of $\Phi_{ij}^{(H)}(\underline{K})$.
Phillips' and our solution both give

$$\overline{u_j^2} = 0.5 \; \overline{u_3^2} \qquad z \geqslant 0^+, \; j=1 \text{ or } 2. \hspace{1cm} (2.12)$$

In order to obtain specific results for the changes in the
variances and spectra near the interface we use two different
expressions for the energy spectrum tensor both with the same
form

$$\Phi_{ij}^{(H)}(\underline{K}) = g_3(k^2\delta_{ij} - K_iK_j)/(g_2 + k^2)^{\mu+2} \hspace{1cm} (2.13)$$

where $\quad k^2 = K_iK_i,$

and with one-dimensional spectra given by

$$\Theta_{ij}^{(H)} (\underline{K}) = g_1 / (g_2 + K_1^2)^\mu \qquad (2.14)$$

In the von Kármán form $\mu = 5/6$, $g_1 = g_2^{5/6}/\pi = 0.1955$

$$g_2 = \pi\Gamma^2(5/6)/\Gamma^2(\tfrac{1}{3}) = 0.558, \; g_3 = \frac{55}{36} g_1/\pi$$

whereas the Townsend form requires $\mu = 1$, $g_1 = 1/\pi$, $g_2 = 1$
and $g_3 = 2g_1/\pi$. The two expressions have both disadvantages
and advantages: the von Kármán spectrum generally gives a more
accurate representation of turbulence, reducing to the
Kolmogorov spectrum $\Theta_{11}^{(H)} (K_1) = g_1/K_1^{5/3}$ for $K_1 \gg g_2$, however
the Townsend spectrum enables the integrals to be calculated
more easily.

2.3   Results for von Kármán and Townsend spectra

2.3.1   One-dimensional spectra:

In the discussion of the results it is convenient to use
the suffices g and e to indicate the contributions to the
spectra from the different frequency ranges:

(i) $\omega < N$ (gravity waves in stratified layers),

(ii) $\omega > N$ (evanescent waves in stratified layer).

Computed spectra using both the Townsend and von Kármán
forms are plotted as functions of z and the non-dimensional
buoyancy frequency NL_H/u_H in Fig. 3. Note how, at z = 0, as
N increases for given u_H, L_H, the value of K_1 for which
$\Theta_{33}(K_1)$ is a maximum also increases, a result of the fact that
the minimum resistance to vertical motion by the stratified
layer occurs when $k_{12} \backsimeq N/u_H$. This resistance increases
rapidly with decreasing k_{12} when $k_{12} < N/u_H$, because the lower
wavenumber, the greater the disturbance would extend into the
stratified region and the greater the energy required to make
such a disturbance. The changes in $\Theta_{33}(K_1)$ as a function of z

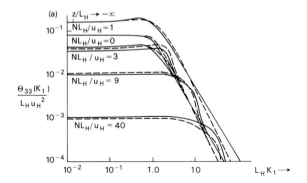

(a)   Normal component at z=0 for different values of NL_H/u_H.

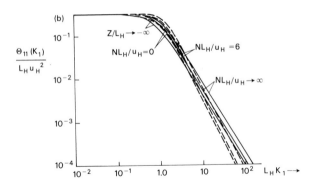

(b)   Longitudinal component at $z=0^-$.

Fig. 3   One-dimensional spectra for two forms of spectrum of
         homogeneous turbulence in the free stream compared with
         the experimental data of Caughey and Palmer (1979)
         where appropriate. _____ von Kármán spectrum;
         ------- Townsend spectrum

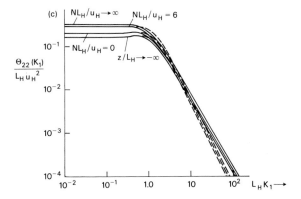

3(c)   Transverse component at $z=0^-$.

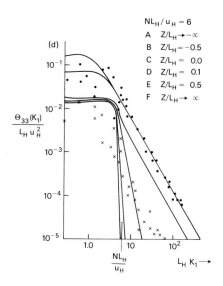

3(d) Normal component for $NL_H/u_H = 6$. Measured
    •, $z = -0.1 \, z_i \, (\approx -0.2L_H)$;   x, $z = 0.4 \, z_i \, (\approx 0.8L_H)$.
    The interface is at $z=0$ in our notation.

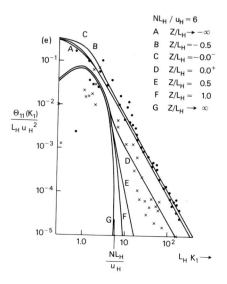

3(e)   Longitudinal component

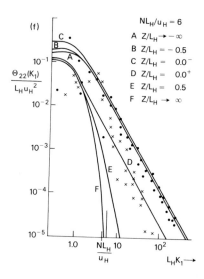

3(f)   Transverse component.

show that there is a rapid fall, off as z increases, in the high-frequency contribution for $K_1 > N/u_H$; this shows that the evanescent term $\theta_{33}^e(K_1)$ decays rapidly with height. At heights greater than about an integral scale above the interface ($z \gtrsim L_H$), the only contributions to the spectra come from gravity waves and these spectra fall off sharply at high frequency. Also in our inviscid model, these spectra do not vary with z. (If, as is likely, absorption occurs at another level, that does not affect the solution near the interface. However reflection does effect this solution as shown in section 3.2.3.) The measurements of Caughey and Palmer (1979) shown in the figures are discussed in detail in section 3.2.

2.3.2 Variances

Analytical expressions can be obtained for the gravity wave contribution for $z \gtrsim 0$, and for the evanescent contribution on z=0.

von Kármán form; $NL_H/u_H \gg 1$:

$$\overline{w_g^2}(z \gtrsim 0) \approx 0.55\, u_H^2 (u_H/NL_H)^{2/3} \tag{2.15}$$

$$\overline{w_e^2}(z=0) \approx 0.32\, u_H^2 (u_H/NL_H)^{2/3} \tag{2.16}$$

Note that the results (2.15), (2.16) depend only on the form of the high-frequency part of the wavenumber-frequency spectrum, which is universal (though unknown), and independent of its form at low wavenumbers. On z=0 (2.15) and (2.16) reduce to

$$\overline{w^2} = \overline{w_g^2} + \overline{w_e^2} \approx 0.87\, u_H^2 (u_H/NL_H)^{2/3} \tag{2.17}$$

The coefficient is determined by the definition of the von Kármán spectrum. Long (1978), in his theory of mixing in stably stratified fluid, finds that at z=0

$$\overline{w^2} \propto u_H^2 (u_H/NL_H) \tag{2.18}$$

This result is given by our theory if a 'Lagrangian' rather than an 'Eulerian' frequency spectrum is used.

In physical terms forcing the interface by a Lagrangian spectrum of the vertical component of turbulence implies, approximately, that waves are produced by eddies rising or falling, whereas forcing the interface by the more energetic Eulerian spectrum implies that waves are largely produced by eddies randomly moving horizontally along the interface. If our expression (2.17) is used to estimate the entrainment rate by using Long's hypotheses, then the entrainment would be increased and be closer to observed values.

It is interesting to note that a similar expression to (2.17) can be obtained on simple physical grounds: assuming for a large enough stratification that the turbulence is a function only of N and of the 'Eulerian' frequency spectrum ($\phi^{(H)}(\omega)$ of the homogeneous turbulence at frequency N, i.e.

$$\overline{w^2} = f(\phi^{(H)}(\omega = N), N) \qquad (2.19a)$$

Then dimensional analysis of this expression leads to

$$\overline{w^2} = c u_H^2 (u_H/NL_H)^{2/3} \qquad (2.19b)$$

where c is a constant and the expression is similar to (2.17).

Also, for $NL_H/u_H \gg 1$,

$$\overline{u_g^2}, \overline{v_g^2} (z \geqslant 0^+) \qquad \frac{1}{2} - 0.55(u_H/NL_H)^{2/3} \qquad (2.20)$$

$$\overline{u_e^2}, \overline{v_e^2}(z=0^+) = 0.10(u_H/NL_H)^{2/3} \qquad (2.21)$$

The computed variances are shown in Fig. 4. Note that the vertical variance increases with NL_H/u_H for small NL_H/u_H with a maximum at $NL_H/u_H \simeq 2$, when the frequency of the energy containing eddies corresponds approximately to N. The von Kármán spectrum results in greater vertical fluctuations for $N \gg 1$ since this spectrum has more energy in the smallest

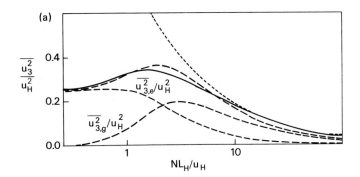

(a)   Normal component at z=0. - - - - - - - - Asymptotic form for
      $NL_H/u_H \gg 1$, von Kármán spectrum.

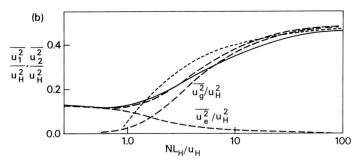

(b)   Longitudinal and transverse components at $z=0^+$.

Fig. 4   Calculated variances for two forms of free-stream
         turbulence compared with the experimental observations
         of Caughey and Palmer (1979) where appropriate.
         _____ von Kármán spectrum   - - - - - - - Townsend
         spectrum.

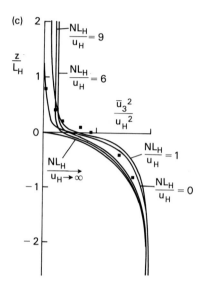

4(c)   Normal component as a function of NL_H/u_H.   Measured:
$NL_H/u_H = 6$, ■.

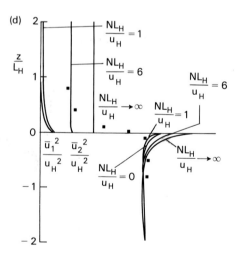

4(d)   Calculated longitudinal/transverse components.   Measured
component $NL_H/u_H = 6$, ■.

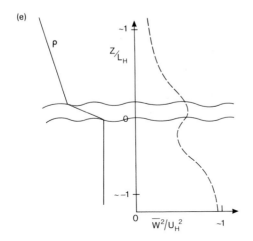

4(e)   Schematic diagram of fluctuating vertical velocity $(\overline{w^2}/u_H^2)$ for case (b).

eddies. Observations from Caughey and Palmer (1979) and numerical calculations from the large eddy simulations of Deardorff (1980) are plotted on the curves where appropriate. These are discussed in Section 3.2. Also plotted in Fig. 4(e) is a schematic profile of the vertical velocity fluctuations from preliminary calculations using the model for case (b). This is also discussed in section 3.2.

Plots of the total kinetic energy are shown in Fig. 5. Note that $\overline{q^2}(z=0^-) < q_H^2$ unless $N \to \infty$. Also, when in dimensional terms $NL_H/u_H \lesssim 3$, q^2 decreases monotonically with increasing z, whereas when $NL_H/u_H > 3$ there is a minimum in q^2 at $z/L_H \approx -0.2$.

2.3.3   Integral scales

These can be calculated analytically on $z=0^-$ for $NL_H/u_H \gg 1$. For the von Kármán spectrum the horizontal scale of the vertical fluctuations

$$L_x^{(w)} \approx 2.7\ u_H/N \qquad\qquad (2.22)$$

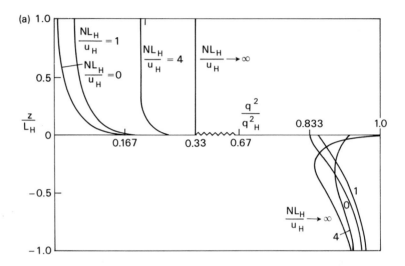

Fig. 5(a)   Kinetic energy (q^2) as a function of z for the von
Kármán spectrum of free-stream turbulence.

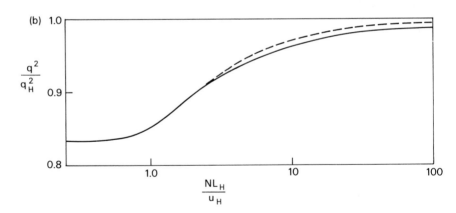

(b)   Turbulence energy at z=0$^-$. _____ von Kármán
spectrum - - - - - - - Townsend spectrum.

Also at z=0

$$(L_x^{(w)}/L_H)/(\overline{w^2}/u_H^2) \propto (u_H/NL_H)^{1/3} \qquad (2.23)$$

and the integral scale decreases at a faster rate with N than the vertical velocity variance. The computed integral scales are plotted as functions of NL_H/u_H on $z=0^-$ in Fig. 6. Only $L_x^{(w)}$ is seen to vary significantly with stratification.

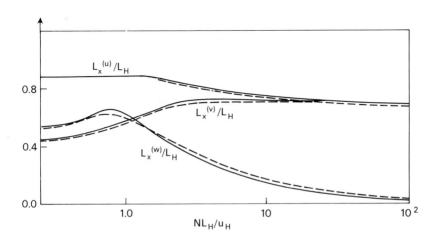

Fig. 6   Integral scales on $z=0^-$ as a function of NL_H/u_H

2.3.4   Wave-energy flux in the stratified region

The vertical flux of wave energy (F_w) can also be calculated in the stratified layer (CH). This flux is independent of z; for $\dfrac{NL_H}{u_H} \gg 1$ using the von Kármán spectrum

$$F_w/\rho u_H^3 \approx 2.2 (u_H/NL_H)^{2/3}, \qquad (2.24)$$

where ρ is the mean density in the layer.

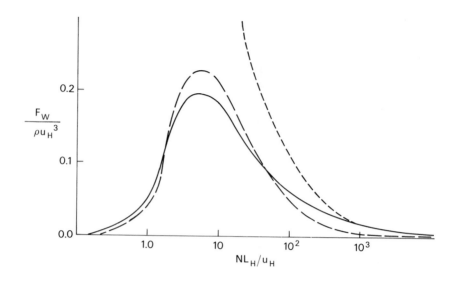

Fig. 7   Calculated vertical wave flux in the stratified
layer _____ von Kármán spectrum ——·——· Townsend
spectrum of free-stream turbulence. ———— Asymptotic form
for $\dfrac{NL_H}{u_H} \gg 1$, von Kármán spectrum.

The computed fluxes and the asymptotic forms for $NL_H/u_H \gg 1$
are shown in Fig. 7. Comparing the rate of energy (power) lost
by the turbulent fluid as wave energy to the energy dissipation
rate in the boundary layer depth z_i

$$\frac{P_{waves}}{P_{dissipation}} = \frac{F_w \bar{\rho} u_H^3}{\epsilon \bar{\rho} z_i} \qquad (2.25)$$

where ϵ is the energy dissipation rate. Taking $z_i = 2L_H$ and
$\epsilon = u_H^3/L_H$ leads to

$$\frac{P_{waves}}{P_{dissipation}} = \frac{F_w}{2} \qquad (2.26)$$

so, since from the computations $F_w \leqslant 0.2$, the loss of energy by
waves is approximately an order of magnitude lower than the
loss by dissipation throughout the depth of the boundary layer.
However the loss of energy by waves does have an important
effect on the dynamics of the mixed layer because it is comparable with the energy dissipated near the inversion; hence there
is less energy available for deepening the mixed layer.

3. DISCUSSION

3.1 Limits of the Model

   In the introduction we suggested that the general class of
turbulent flows confined by stratification can be subdivided
into three main types (Fig. 1): (a) the stable layer is
uniformly stratified and there is no mean shear; (b) there is
an intensely stratified layer bounding the edge of the turbulent layer, with no shear; (c) either of the first two cases
with mean shear. The theory described in this paper and in CH
is a model of type (a). A further modification of the analysis
is to take account of an intensely stratified layer marking
the top of the boundary layer (type b) will be described in a
subsequent paper; however important additional physical effects
due to the presence of the intensely stratified layer will be
described in section 3.2.

   The theory does not include the effect in the measured
spectra and variances of the undulating motions at the interface caused by the eddies impinging onto it. The effect of
this on time-averaged measurements at a point is to blur the
discontinuity in the structure of the horizontal components of
turbulence across the interface. Gartshore, Durbin and Hunt
(1983), in their use of rapid distortion theory to describe the
structure of motions in a shear layer outside a turbulent
layer, estimated the effect by assuming the position of the
interface deviated from its mean portion according to a
Gaussian probability distribution.

   Waves impinging on the interface from aloft are not considered by the theory. These have been described by Delisi
and Orlanski (1975). They found that there was a maximum in
wave amplitude near the interface where incident and reflected
waves interact. The amplitudes were small unless there was a
thin region of stronger stratification marking the edge of the
neutral layer; in this case wave breaking was observed to occur
and large-amplitude waves were predicted by their theory.

3.2 Comparisons with experiment and numerical simulations

Detailed observed and numerically simulated structures of
turbulence near an interface between a turbulent layer and a
single uniformly stratified layer are rare, and the range of
the nondimensional buoyancy frequency is small; typically in
the atmosphere $NL_H/u_H \approx$ 6-8. Specific comparisons are made
with the balloon measurements of Caughey and Palmer (1979) made
in the dry atmosphere boundary layer, the laboratory experiment
of Willis and Deardorff (1974) and the numerical simulations
of Deardorff (1980), (Cases 1,2 and 3).

We also refer to the more numerous observations of turbulence confined by a strongly stratified layer with a less
stratified (or neutral) layer beyond. These include the
doppler radar measurements of Gossard et al. (1982), the
balloon experiments of Caughey et al. (1982) (in stratocumulus
clouds), the laboratory tank experiments of McDougall (1979),
and Fernado and Long (1983), and further numerical simulations
of Deardorff (1980).

3.2.1 Conditions of the theory

(i) Energy dissipation rate

The theory requires $\partial \varepsilon/\partial z = 0$ or more precisely,

$$\frac{1}{\varepsilon}\frac{\partial \varepsilon}{\partial z} \ll (L_x^{(w)})^{-1} \qquad (3.1)$$

(Hunt, 1984). This condition is well satisfied by the simulations of Deardorff where a balance of the turbulent kinetic
energy equation requires $\frac{\partial \varepsilon}{\partial z} \approx 0$. In Caughey and Palmer
$\frac{1}{\varepsilon}\left|\frac{\partial \varepsilon}{\partial z}\right| \approx \frac{1}{2L_H}$. With a strongly stratified layer at the top of
the turbulent region, there is usually an increase in ε at the
top of the turbulent layer.

(ii) Anisotropy

The measurements of Caughey and Palmer show that the
variances of the turbulent components in the centre of the
boundary layer are approximately equal

$$\overline{w^2} \approx \overline{u^2} \approx 0.4\, w_*^2 \qquad (3.2)$$

where $w_*^2 = g \dfrac{\overline{w'\theta'}\big|_0 \cdot z_i}{\overline{\theta}}, \overline{w'\theta'}\big|_0$ is the surface heat flux,

and z_i the depth of the boundary layer. This does not mean that the turbulence is isotropic; in fact boundary-layer convection usually consists of energetic upward motions of relatively small extent penetrating larger regions of slowly descending air (Lenschow and Stephens, 1980), which is why the third-order moments $(\overline{w^3})$ are positive.

The computer simulations of Deardorff and laboratory experiments of Willis and Deardorff showed a highly anisotropic turbulent structure with

$$\frac{\overline{w^2}}{w_*^2} \approx 0.5, \quad \frac{\overline{u^2}}{w_*^2} \approx 0.2. \tag{3.3}$$

But in these cases $\overline{w^3}/w_*^3$ was comparable with atmospheric observations and was large and positive throughout most of the depth of the boundary layer.

However the atmospheric measurements do show the small scale isotropy and universal spectral form assumed in this theory and that of Hunt (1984).

(iii) Temperature/density gradient in the neutral layer

The theory assumes that the turbulent region is sufficiently well mixed that there is no significant stratification there. However entrainment of fluid from the stable layer is observed to cause low levels of stratification in the upper half of the turbulent region with, for the experiments and simulations referred to, buoyancy frequency $N \lesssim 4 \times 10^{-3}\,\text{sec}^{-1}$. This stratification may be expected to have a dynamic effect on scales L for which $u_H/NL \lesssim 1$. This inequality is only satisfied for scales larger than the scale over which the temperature gradient occurs, so dynamic effects are likely to be small. The nonlinear effects due to the interaction of eddies and fluctuating temperature gradients near the surface due to the heat source are discussed by Hunt (1984).

3.2.3 Comparison

(i) Spectra

The measured spectra of Caughey and Palmer are compared with
the predictions of the theory in Figs. 3d, e and f. The non-
dimensional buoyancy frequency $NL_H/u_H = 6$, where we have taken
$L_H = z_i/2$ and z_i is the depth of the boundary layer (z=0 at
the boundary layer top in our notation and the earth's surface
is at $z=-z_i$). The detailed temperature and velocity profiles
are given by Palmer et al. (1979). The curves are in qualitative agreement, with the sharp fall-offs in the high frequency
($K_1 \gg N/u_H$) parts of the spectra for $z > z_i$; the residual
high-frequency contribution in the measurements at $z = 0.4z_i$
is probably due to instrumental noise. The spectra of Gossard
(private communication) calculated from Doppler radar measurements also show the same features in a stable layer well above
the interface (although in this case there was a narrow intense
stable layer at the top of the boundary layer, in the deep
stable layer above this, the theory of this paper is
appropriate).

(ii)   Variances

The marked anisotropy of the boundary-layer turbulence in
the experiments of Willis and Deardorff and in the numerical
simulations of Deardorff makes precise comparison with the
theory difficult; however in all cases the observations show
broad agreement with the theory in the sense that the vertical
variances decrease from their maximum values near the centre
of the boundary layer as the interface is approached. The
horizontal components show little variation over the entire
mixed layer depth, but there is some evidence of a small
increase near the interface. Figs. 4c and 4d show the measured
values of Caughey and Palmer compared with the theory. The
smooth profile in the time-averaged horizontal variance across
the interface is caused by the undulating motions of the
interface. Table 1 shows calculated and measured values of
$\overline{w^2}/u_H^2$ at the interface. It is assumed that $u_H = W_0$ where W_0
is the r.m.s. value of the observed vertical velocity at the
centre of the mixed layer.

It is interesting that when a strongly stratified layer
caps the boundary layer with a less stable layer beyond (case
(b) in the Introduction), the numerical simulation of Deardorff
shows secondary maxima in $\overline{w^2}$ in the strongly stratified layer

Table 1

Vertical variances. Comparisons of the theory with the experiments of Willis and Deardorff and the numerical simulations of Deardorff.

	$\dfrac{NL_H}{u_H}$	$\overline{w}^2/w_O^2$ (z=0) observation/ simulation	Theory	$\overline{w}^2/w_O^2$ (z=0.1 L_H) observation/ simulation	Theory
Willis and Deardorff Experiment					
1	7.0	0.26	0.21	0.19	0.16
2	10.0	0.18	0.16	0.09	0.11
Deardorff numerical simulation					
1	8.0	0.2	0.23		
2	7.1	0.3	0.24		
3	8.6	0.32	0.22		

at the top of the boundary layer. A likely cause is resonance
and the growth of trapped waves in the very stable layer (e.g.
laboratory experiments of Fernando and Long (1983), and the
atmospheric observations of Caughey et al. (1982). Preliminary
results from a three-layer model for case (b) (Fig. 4(e)) show
the similar variance profiles to those observed including the
second maximum in $\overline{w^2}$. This model uses the assumptions of the
two-layer model described in this paper and a plausable
Richardson number criterion to limit the growth of the trapped
waves. The eventual breaking of the trapped waves is one of
the entrainment mechanisms described in the Introduction. In
the less stable region well above the boundary layer, three
and two-layer models exhibit similar behaviour: the high-
frequency components rapidly decay with height while the low-
frequency components propagate as gravity waves.

(iii)   Other computational methods

   There have been numerous attempts to model the velocity and
temperature variances in the upper part of the convective
boundary layer using ensemble averaged momentum and conservation equations. First and second order closure gradient
transport schemes can not transport heat and energy against the
mean temperature and turbulence gradients (Zeman, 1975) so
that most recent modelling has used the third order equations
(André et al., 1976; Lumley, Zeman and Siess, 1978). André
et al. used a quasi-normal approximation for the transport
equations together with a realizability condition (Wyngaard,
1982) to ensure positive values of energy, whilst Lumley et al.
also used a quasi-normal approximation and neglected pressure
correlation terms. The predictions of both models were
compared with the tank experiments of Willis and Deardorff
(1974). In the interior of the turbulent region calculated
variances and turbulent transport terms are similar to experimental values. Near the interface the closure models overestimate both $\overline{w^2}$ and the third order terms, the error probably
occurring because these models cannot treat wave motion
separately from turbulent motion as is possible using the
linear theory described in this paper. Well above the interface the closure models damp out all motion and all fluxes,
whereas the observations of Caughey et al. and the linear
theory show that wave motion can extend well above the interface and can transfer a significant flux of energy.

3.3 Conclusions

   The linear theory describes the main features observed near
an interface between a turbulent region and a stably stratified

layer. For case (a) these include a decrease in $\overline{w^2}$ near the interface and the propagation of the gravity waves into the stratified layer with a frequency ω less than the buoyancy frequency. A prediction of the theory is that $\overline{w^2}$ is a maximum at the interface (z=0) when $NL_H/u_H \approx 1$ while

$$\overline{w^2}(z=0) \propto u_H^2 (u_H/L_H N)^{2/3} \text{ for } NL_H/u_H \gg 1.$$ The theory also predicts that the eddies which penetrate most easily into the stratified layer are those with frequency $\omega = N$. It is likely that these eddies are important in the entrainment process (mechanism iii).

For case (b), $\overline{w^2}$ exhibits a second maximum in the strongly stratified layer marking the top of the boundary layer where waves of frequency $\omega > N$ are trapped. In this case N is the buoyancy frequency of the upper layer.

For the above turbulent region the models for both cases (a) and (b) show similar behaviour with propagating waves of frequency $\omega < N$, but little or no energy at frequency $\omega > N$. Wavenumber spectra show very sharp fall-offs at wavenumbers corresponding to frequency N.

ACKNOWLEDGEMENTS

We are grateful for stimulating conversations with N.H. Thomas of the Chemical Engineering Department, Birmingham University and J.W. Rottman of DAMTP, Cambridge University. D.J.C. was in receipt of an NERC fellowship during the course of this work.

REFERENCES

André, J.C., DeMoor, G., Lacarrere, P. and Du Vachat, R. (1976) Turbulence approximations for Inhomogeneous Flows, *J. Atmos. Sci.*, **33**, 476-491.

Brost, R.A., Wyngaard, J.C. and Lenschow, A.H. (1982) Marine stratocumulus layers. Part II. Turbulence budgets. *J. Atmos. Sci.*, **39**, 818-837.

Carruthers, D.J. and Hunt, J.C.R. (1984) Velocity fluctuations near an interface between a turbulent region and a stably stratified layer. Submitted to *J. Fluid Mech.*

Caughey, S.J., Crease, B.A. and Roach, W.T. (1982) A field study of nocturnal stratocumulus. II. *Quart. J. Roy. Met. Soc.*, **108**, 125-144.

Caughey, S.J. and Palmer, S.G. (1979) Some aspects of
    turbulence through the depth of the convective boundary
    layer. *Quart. J. Roy. Met. Soc.,* **105**, 811-827.

Deardorff, J.W. (1980) Stratocumulus-capped mixed layers
    derived from a three-dimensional model. *Boundary Layer
    Met.,* **18**, 495-527.

Delisi, D.P. and Orlanski, I. (1975) On the role of density
    jumps in the reflection and breaking of internal gravity
    waves. *J. Fluid Mech.,* **69**, 445-464.

Fernando, H.J.S. and Long, R.R. The growth of a grid
    generated turbulent mixed layer in a two-fluid system.
    J. Fluid Mech., **133**, 377-396.

Gartshore, I.S., Durbin, P.A. and Hunt, J.C.R. (1983) The
    production of turbulent stress in a shear flow by irrota-
    tional fluctuations. *J. Fluid Mech.,* **137**, 307-330.

Gossard, E.E., Chadwick, R.B., Neff, W.D. and Moran, K.P.
    (1982) The use of ground-based doppler radars to measure
    gradients, fluxes and structure parameters in elevated
    layers. *J. App. Met.,* **21**, 211-226.

Hunt, J.C.R. (1984) Turbulence structure in thermal convection
    and shear-free boundary layers. *J. Fluid Mech.,* **138**,
    161-184.

Hunt, J.C.R. and Graham, J.M.R. (1978) Free stream turbulence
    near plane boundaries. *J. Fluid Mech.,* **84**, 209-235.

Hunt, J.C.R., Rottman, J.W. and Britter, R.E. (1984) Some
    physical processes involved in the dispersion of dense
    gases. Proc. IUTAM Symposium on Atmospheric dispersion of
    heavy gases and small particles. Delft, The Netherlands,
    August 1983.

Kaimal, J.C., Wyngaard, J.C., Haugen, D.A., Coté, O.R.,
    Izumi, Y., Caughey, S.J. and Readings, C.J. (1976) Turbulence
    structure in the convecture boundary layer. *J. Atmos. Sci.,*
    33, 452-2169.

Kato, A. and Phillips, O.M. (1969) On the penetration of a
    turbulent layer into a stratified liquid. *J. Fluid Mech.,*
    37, 643-655.

Kovasnay, L.S.G., Kibens, V. and Blackwelder, R.F. (1970)
    Large scale motion in the intermittent region of a turbulent
    boundary layer. *J. Fluid Mech.,* **41**, 283-325.

Lenschow, D.H. and Stephens, P.L. (1980) *Boundary Layer Met.,*
   19, 509.

Linden, P.F. (1973) The interaction of a vortex ring with a
   sharp density interface: a model for turbulent entrainment.
   J. Fluid Mech., **60**, 467-480.

Long, R.R. (1978) A theory of mixing in stably stratified
   fluid. *J. Fluid Mech.,* **84**, 113-124.

Lumley, J.L., Zeman, O. and Siess, J. (1978) The influence
   of buoyancy on turbulent transport. *J. Fluid Mech.,* **84**,
   581-597.

McDougall, T.J. (1979) Measurements of turbulence in a zero
   mean-shear mixed layer. *J. Fluid Mech.,* **954**, 409-431.

Palmer, S.G., Caughey, S.J. and Whyte, K.W. (1979) An
   observational study of entraining convection using balloon-
   borne turbulence probes and high power Doppler radar.
   Boundary Layer Met., **16**, 261-278.

Phillips, O.M. (1955) The irrotational motion outside a free
   boundary layer. *Proc. Camb. Phil. Soc.,* **51**, 220.

Piat, J.F. and Hopfinger, E.J. A boundary layer trapped
   by a density interface. *J. Fluid Mech.,* **113**, 411-432.

Price, J.F. (1979) On the scaling of stress-driven entrainment
   experiments. *J. Fluid Mech.,* **90**, 509-529.

Tennekes, H. (1975) Eulerian and Lagrangian time microscales
   in isotropic turbulence. *J. Fluid Mech.,* **67**, 561-567.

Tennekes, H. and Lumley, J.L. (1972) A short course in
   turbulence. MIT Press, 293.

Townsend, A.A. (1966) Internal waves produced by a convective
   layer. *J. Fluid Mech.,* **24**, 307-319.

Turner, J.S. (1973) Boundary Effects in Fluids. Cambridge
   University Press, 368.

Willis, G.E. and Deardorff, J.W. (1974) A laboratory model of
   the unstable planetary boundary layer. *J. Atmos. Sci.,* **31**,
   1297-1307.

Wyngaard, J.C. (1982) Boundary layer modelling. Atmospheric
    turbulence and air pollution modelling. D. Reidel Publ.,
    69-106.

Zeman, O. (1975) The dynamics of entrainment in the planetary
    boundary layer: a study in turbulence modelling and para-
    meterization. Ph.D. Thesis. The Pennsylvania State
    University.

MIXING OF GRAVITY CURRENTS IN TURBULENT SURROUNDINGS: LABORATORY STUDIES AND MODELLING IMPLICATIONS†

N.H. Thomas* and J.E. Simpson

(DAMTP, University of Cambridge)

ABSTRACT

A saline gravity current introduced at the downstream end of a water channel travels along the bottom beneath a fresh streaming counter flow. Mean shear in the stream is kept small by using a moving belt floor upstream of the gravity current; the belt travels at the depth-mean speed. Turbulence in the stream is maintained by a horizontal square mesh grid which covers the entire planform area of the channel and is forced to execute vertical oscillations near the free surface. The amplitude of the oscillations is small compared with the depth of flow and the grid remains submerged at all times. In effect, the grid acts as a kind of 'dynamic roughness' on the streaming flow. Salt solution is continuously supplied at the downstream end of the gravity current to compensate for fluid transported into the streaming flow by the action of interfacial shear and turbulence. The supply is adjusted until the front of the gravity current stablises at the leading edge of a fixed floor which extends to the downstream end of the test section.

Experimental measurements are reported for the saline flux - i.e. the time-mean interfacial exchange (mixing) flux averaged over the length of the current - and its dependence on the mean speed and rms turbulence of the streaming flow, and on the length and buoyancy of the gravity current. Only at very low

* Present address: Department of Chemical Engineering, University of Birmingham, P.O. Box 363, Edgbaston, Birmingham, B15 2TT.

† This work was supported (1978-81) by the Central Electricity Generating Board - Central Electricity Research Laboratories, Leatherhead.

turbulence intensities, less than 1/2 % or so, do our results
approach those obtained by Britter and Simpson (1978) for non-
turbulent flows in the same equipment. Under these conditions,
the transport is predominantly due to an interfacial shear
instability near the front of the gravity current. The interface is thickened by these overturning and mixing motions, and
is thereby stabilised against disturbances in the downstream
flows. On the other hand, at very high turbulence intensities,
greater than 50% or so, our findings are consistent with those
obtained by Turner (1968) in experiments on turbulent layers in
a static tank; Turner also used an oscillating grid to drive
the turbulence. The transport in this case is maintained by
the entraining eddies as they impinge on the interface. For
the condition of our experiment, i.e. interfacial turbulent
Richardson numbers in excess of about 10, the process is always
more akin to erosion rather than turbulent diffusion. That is
to say, the flux occurs without significant evolution in either
the interfacial density gradient or the salinity of the non-
turbulent gravity current.

1.   INTRODUCTION

   Many experimental studies have been concerned with the
fundamentals of gravity currents in non-turbulent streams or
gravity layers in non-streaming turbulence. Two examples are
highlighted here because of their particular relevance to the
present experimental investigation into the mixing behaviour
of gravity currents in turbulent streams. Firstly, Britter
and Simpson (1978-BS below) who used a mini channel with moving
belt floor to sustain shear-free steady streaming flow and,
secondly, Turner (1968 - T below) who used an oscillating grid
to sustain non-streaming turbulence in a rectangular box. The
apparatus we employed is simply a composite of these two items
of equipment. Some main elements of the earlier work by BS and
T and other relevant studies are recapped below, followed by
an outline of this paper and some practical relevancies of our
findings.

1.1 Gravity currents in non-turbulent streams

(1) Apparatus The apparatus used by BS is sketched on
Fig. 1(a); see their paper for details and dimensions. The
streaming flow (with speeds up to about 10cm/s) is maintained
in a channel of bench top scale (15cm or so square cross-section
and about 75cm long). Vertical shear is kept small by a moving
belt floor travelling at the depth mean speed of the flow.
A saline gravity current introduced at the downstream end
travels counter to the streaming flow along a fixed plate floor
in the working section of the channel. The densities ρ_s of the

stream and ρ_c of the current are maintained by reservoir
supplies and the interfacial buoyancy is well approximated by
the Boussinesq reduced gravity parameter $\Delta g = g(\rho_c - \rho_s)/\rho_s$;
c and s denote current and stream quantities throughout our
paper. With Δg prescribed, the stream speed U_s and the current
flux Q_c are adjusted to achieve the equilibrium portrayed in
the sketch, i.e. with the frontal edge of the current located
at the leading edge of the fixed plate. The arrangement thus
provides for a stationary model of a gravity current propagating in non-turbulent surroundings.

    The frontal conditions of the current are not (quite)
uniquely determined in the experiment. That is to say, there
is a range of flow speeds and current fluxes for which the
front of the current settles at the leading edge of the plate,
corresponding to a range of interfacial angles θ of the frontal
edge; see Fig 1(a). This contrasts with the theoretically
ideal (i.e. inviscid and non-mixing) analysis of Benjamin
(1968) whose solution insisted that $\theta = \pi/3$. In practice, U_s
and Q_c are not so sensitive to the range of equilibria usually
attained in the experiments ($\pi/6 \leqslant \theta \leqslant \pi/3$, say), so this
ambiguity is inconsequential, to a first approximation at
least. Indeed BS obtained respectably repeatable measurements
of Q_c and its dependence on U_s, Δg and H_c , the depth of the
current (see Fig. 1(a) and below), as well as data on the
dynamics of the gravity current flows (see their paper).
Importantly, they set aside doubts about Reynolds number
distortion in such a small bench top model, by demonstrating
that Q_c is insensitive to the internal current Reynolds number
(Q_c/ν_c where ν_c = kinematic viscosity) for typical conditions of
their experiments.

(2) Results A tracing of one of the shadowgraph visualisation
photographs published by BS appears as Fig. 1(b) here. It
nicely illustrates how the mixing structure is dominated by
large scale instabilities localised in the sharp interface at
the front of the current. These Kelvin-Helmholtz waves first
grow then overturn and roll up, engulfing fluid from both the
current and the stream until they eventually collapse to leave
a shear-mixed interfacial layer whose vertical structure is
stable to downstream shear disturbances. In support of this
picture of the events - idealised in the sketch of Fig. 1 (a),

(a)

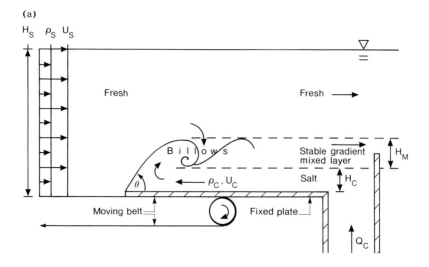

Fig. 1(a)   Sketch of the apparatus used by Britter and Simpson
            (1978) for studies of gravity currents in non-
            turbulent streams.   Mean shear is suppressed by the
            moving belt floor.

(b)

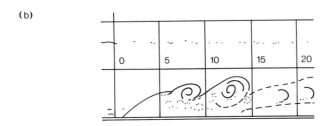

Fig. 1(b)   Tracing of shadowgraph image taken from photograph
            published as Fig. 1(b) of Britter and Simpson.   The
            mixing structure is dominated by large scale shear
            instabilities confined to the nose of the current.

BS demonstrated that the mixing flux is adequately described by
simple empirical   formulae like

$$\Delta g \, Q_c / U_s^3 \simeq 0.15 + \text{weak function} \; (H_c / H_s) \qquad (1.1a)$$

and

$$\Delta g \; H_m / U_s^2 \simeq \text{constant} . \qquad\qquad (1.1b)$$

H_s is the stream depth and H_m the depth of the stable mixed
layer. Both of these results are consistent with ideas based
on elementary energetics of localised interfacial mixing, i.e.
that only a small proportion (typically about 1/5 or so) of the
available shear kinetic energy at the interface is converted to
potential energy of mixing, the remainder being lost to
dissipation; e.g. Linden's (1979) data review. This finding
has also been deduced from an elementary first principles
analysis utilising all of the flow invariants for concurrent
mixing streams (to be published).

1.2 Gravity layers in non-streaming turbulence

(1) Apparatus A sketch appears as Fig. 2(a); see T for
details of construction and operation. The square mesh grid
(5cm meshes constructed from 1cm bars) is mounted horizontally
and executes vertical oscillations (up to about 10 Hz, with
1cm stroke typically) to supply turbulence which fills the
stirred liquid in the static tank (about 25cm square planform
and 40cm deep). In the absence of any local measurements of
the velocity statistics at the time of his study, Turner
ignored any mean flows and accordingly assumed the turbulence
to be nominally homogeneous in horizontal planes parallel to
the grid. He supposed power law dependences on distance z'
measured from the grid mean plane for both the decay in rms
fluctuations U' and the growth in eddy size L' (integral
scale, say). Since then, however, detailed measurements by
Hopfinger and Toly (1976) and McDougall (1979) have revealed
strong mean flows (bulk swirl) which depend on the frequency
of oscillation and are sensitive to fine details of the
geometry - for example, the clearance between the bars of the
grid and the side walls of the tank. Grid submergences of less
than 1 mesh length or so can also introduce similar compli-
cations (see sections 2,3). Nevertheless, these later studies
have also demonstrated that there is a useful operating range
of frequencies for which the mean flows can be reasonably
neglected. Moreover, that within this range, the turbulence
quantitites U' and L', which depend on a mesh size M, bar size
d, stroke S , frequency f, as well as z', are indeed adequately
represented by power law dependences on z'. Thus Hopfinger
and Toly proposed the following as correlation formulae for

(a)

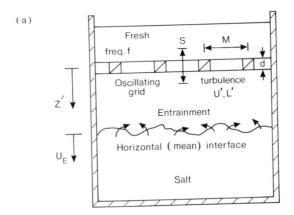

Fig. 2(a)   Sketch of the apparatus used by Turner (1968) for
            experiments on mixing of gravity layers in non-
            streaming turbulence.

(b)

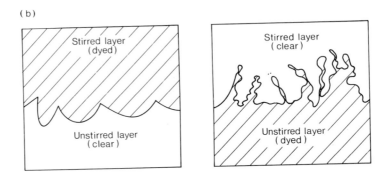

Fig. 2(b)   Sketches to illustrate physical features of the
            mixing interface, particularly the convoluted thin
            shear layer (Hopfinger and Toly, 1976) and the wispy
            tails of entrailed fluid (Turner).

their data:

$$U'/fS \simeq \tfrac{1}{4} \sqrt{Md}/z' \quad (\pm 20\%)$$

$$L' \simeq (0.2 \pm 0.1)z'$$

$$(1.2a)$$

the latter depending in an unclear fashion on M and S, and
providing

$$1 \leqslant S/d \leqslant 4, \ M/d \approx 5, \ 2 \leqslant f(Hz) \leqslant 6$$

and   (1.2b)

$$z'/M \geqslant 3/2$$

as required for horizontal homogeneity.

No measurements of the turbulence quantities are undertaken
in our study reported below and our data analysis makes use of
these correlations throughout.

(2) Results The experimental procedure followed in T was
simple and straightforward. The tank was primed with two
layers, fresh above saline (or heated above cooled) and
separated by a sharp interface. The grid was switched on and
the interface was visualised either by shadowgraph or by a dye
discriminator in one or other of the layers. The sketches of
Fig. 2(b) here illustrate the important physical features
brought to light by Turner's published photographs. For
sufficiently strong buoyancy between the layers (i.e.

$R_i' = \Delta g L'/U'^2 \geqslant 5$ or so - see section 4), the initial interface
remains sharp or is further sharpened, behaviour which is in
striking contrast to the smearing and thickening of non-
buoyant interfaces (see Crapper and Linden, 1974). The
convoluted and wispy boundary of the entraining buoyant layer
suggests a mechanism akin to erosion by the eddies as they
impinge on, and are arrested by, the stable interface; cf.
turbulent 'diffusion' of material dispersed by quasi-neutral
eddies travelling across an 'interface' into non-turbulent
surroundings.

Turner made good use of the persistence and sharpness of the
interface to obtain quantitative estimates of the entrainment
speed U_E from a time record of its motion away from the
oscillating grid. These results were confirmed by using a
conductivity probe to measure the evolution of the approximately uniform concentration of dissolved salt (or temperature)
in the stirred layer.   He argued U_E must be a function of the
effective values of U' and L' near the interface, Δg across the
interface and the molecular transport properties ν and κ,
viscosity and diffusivity of the buoyancy agent (i.e. salt or
heat). For large $R_i' (\gtrsim 20)$, and with $R_e' = U'L'/\nu$ in the range

$10 \lesssim R'_e \lesssim 30$, Turner felt his results were consistent with the following formulae:

$$U_E/U' \propto R'^{-3/2}_i \text{ for salt } (P_r \simeq 500)$$

and

$$U_E/U' \propto R'^{-1}_i \text{ for heat } (P_r \simeq 5)$$

where $P_r = \nu/\kappa$ is the Prandtl number. He thence argued that U_E/U' depends only on R'_i and $P'_e (= U'L'/\kappa$ - the Peclet number), but not R'_e . His conjectures about the precise functional form of this relation at large P'_e (i.e. with R'_i exponent equal to -3/2) have been disputed by Hopfinger and Toly, who also put forward their own proposal (with same R'_i exponent) for a data correlation to encompass both Turner's findings and their own results obtained at much larger R'_e, roughly $200 \lesssim R'_e \lesssim 700$. On the other hand, McDougall's more recent experiments and his reexamination of Turner's data led him to suggest a smaller magnitude value for the exponent of R'_i, about -1.2 instead of -3/2.

Further consideration of the power law formulae will be offered in section 3, because our own findings also bear on whether (and if so, how) the entrainment speed may depend on R'_e. Although discounted in earlier work, the matter is of practical importance in assessing scale distortion in the small models used for laboratory experiments.

1.3 *Outline of paper and practical relevance of findings*

The present experiment is reported in section 2. Shadow-graph visualization is used to demonstrate how the large-scale overturning structure of the non-turbulent interfacial shear instability gives way to the wispy-convoluted interface typical of erosive entrainment by turbulent eddies in the absence of mean shear. Our results for the mixing flux show how it becomes insensitive to the turbulence intensity at low intensities and insensitive to the mean speed of the stream at high intensities. For the latter conditions we demonstrate that the mixing flux is approximately proportional to the length of the current. This finding is consistent with our interpretation that eddy transport sustains the mixing flux, with entrainment speed approximately uniform along the length

of the current. We confirm the existence of a power law
relationship between the entrainment speed normalized on rms
velocity fluctuations and the interfacial turbulence Richardson
number. The exponent of our power law is comparable with that
first inferred by Turner for interfacial transport in stirred
turbulent layers. However, the value of our proportionality
coefficient is significantly larger (between two and five
times) than found in previous studies.

This anomaly is investigated in section 3. Our flow channel
was modified to allow measurements of the mixing flux across
a horizontal interface in approximately zero-mean flow,
thereby reproducing conditions found in the static tank
experiments. These data were comparable with our earlier
findings, so we attribute the anomaly to interference effects
between the grid-driven flows and wave motions excited at the
nearby free surface. The phenomenon has since been confirmed
and has been the subject of recent investigations by Dr. Jia
Fu, a visitor to DAMTP; preliminary results of his study are
reported by Jia Fu and Thomas (1983).

In section 4 we put forward some elements for simple
mathematical modelling to characterise the composite effects
of the main mixing factors, viz the shear mixing at low
intensities and the eddy entrainment at high intensities.
These findings should be directly relevant to the effect of
environmental turbulence on hot-water plumes discharged into
natural waters from power stations and accidental releases
of dense gases into the atmosphere. Our findings indicate
that the turbulence may not significantly affect the surface
excess temperatures of hot water plumes, nor the ground level
concentrations of dense gas releases, unless the interfacial
turbulence Richardson number is less than about three or so.

Our modelling approach may be able to encompass other effects
arising from, but not explicitly addressed in, the present
experimental study. For example, we argue that due account
should be taken of the attenuation of mixing arising from
decay in the turbulence as it flows over the shear-relaxed
interface of a gravity current travelling on the shear-
turbulent boundary. This consideration may be important for
the mixing of dense gas releases. We also note how the
spreading of a three dimensional buoyant plume enhances the
contribution of interfacial turbulent mixing as compared with
the shear mixing which is confined to the edges of the plume.
One implication is that the far field mixing of buoyant hot
water discharges from power stations may often be dominated by
detrainment of plume fluid into the streaming flow as the
interface of the heated surface layer is eroded by the external

turbulence. In both of these practical situations the exchange
flux can be maintained without dilution of the boundary
concentration of the buoyant contaminant.

2. GRAVITY CURRENTS IN TURBULENT STREAMS

2.1 Experimental set-up

(1) <u>Apparatus</u> A sketch of our apparatus appears on Fig. 3(a)
and a perspective view is shown in the photograph Fig. 3(b).
The mini-channel with its moving belt floor and fixed plate
working section are as outlined in section 1. The turbulence
generated by the grid propagates downwards whilst being advected by the stream, until it fills the entire depth of flow
after an initial development length at the upstream end of the
channel. Stream conditions approaching the test section
containing the gravity current were approximately fully-
developed for the high intensity flows of main interest here.

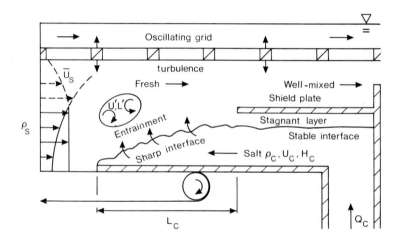

Fig. 3(a) Sketch of the apparatus used for our studies of
           gravity currents in turbulent streams. The action
           of the oscillating grid on the streaming flow is
           akin to 'dynamic roughness'. Unrepresentative
           mixing of the entering current flow is suppressed
           by the shield plate.

   One modification to BS arrangement (Fig. 1) is the shield
plate   mounted horizontally above the saline water duct at
the downstream end of the channel. This was introduced to

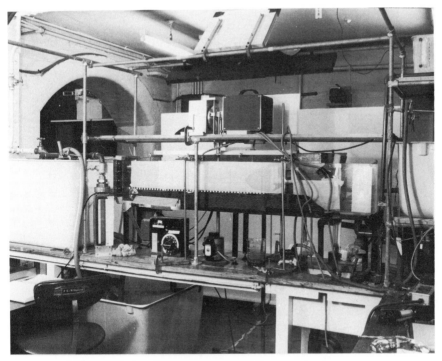

Fig 3(b)   Photograph of the apparatus

protect the entering flow against unrepresentative mixing
during its transition from vertical upwelling at the duct exit
to a sensibly horizontal gravity current in the working section
of the channel.   Shadowgraph observations and measurements of
the mixing flux confirmed that the plate had no discernible
effects on the behaviour of currents in non-turbulent streams.
In high intensity turbulence ($U'/\bar{U} \gtrsim 20\%$, say - where the prime
denotes rms and the bar time mean values), the fluctuating
incidence of the stream approaching the sharp leading edge of
the shield plate gives rise to localised flow separation.   The
weak eddying motions here caused some entrainment of the
current fluid, but this was always small compared with the
total mixing flux.   Experiments were undertaken with different
lengths L_c of gravity current exposed to stream turbulence.

A sliding extension to the fixed plate section of floor allowed
us to adjust the supported length of current fluid as desired.

   The dimensions of our oscillating grid (M and d comparable
with those used in T) were chosen to match those of the planform geometry of the channel - with allowances for small

clearance gaps between the bars of the grid and the side walls
of the channel. The corresponding ratio, length-to-width, of
the grid was about 5:1 and this is well-removed from the square
planform used in all previous work on oscillating grids in
static tanks. Thus, there is some uncertainty about the
validity of the correlation formulae (1.2a) for U' and L'
which we adopted, without verification, in the present study.
Although the results indicate that our assumption may perhaps
have been justified, further turbulence measurements, in
configurations like ours, would clearly be desirable. More
serious uncertainties about the applicability of the correlation formulae arise from the very shallow submergence
($\frac{1}{2}$ M or so) of the grid in our experiments as compared with
the deeply submerged (many M) grids in previous work on static
tanks. The siting of our grid was imposed by geometrical
constraints on the available depth of streaming flow (about
12cm or $\frac{5}{2}$M) and the need to maintain a reasonable minimum depth
of entering current flow (about 2cm or $\frac{1}{2}$M say), together with
the requirement for a spacing of at least $\frac{3}{2}$M between the grid
and the interface, as demanded for horizontal homogeneity of
the turbulence, according to correlation formula (1.2b). As
we shall see from the results presented below and as confirmed
in other recent studies cited in sections 3 and 4, it turns out
that shallow submergence of the grid certainly does affect the
validity of the correlation formula for U'.

On a more general note, it seems to us that the oscillating
grid in streaming flows might be viewed as a 'dynamic roughness' device. As such, it could find wider laboratory
application for simulating the turbulence of wall-sheared flows
in bench top models, perhaps rather like Coles' (1974)
synthetic turbulent boundary layer generated by randomly pulsed
wall jets. For reasons of convenience (and economics),
laboratory models are often of such small scale that the
attainable bulk Reynolds numbers, or the entry lengths allowed
for flow development, may be insufficient to adequately
represent the levels and distributions of turbulence found in
prototype or field situations. We believe that wall roughness
with dynamic parameters S and f to augment the static parameters M and d could offer considerable scope for investigating
and estimating scale distortions like these.

(2) Procedure The procedure we followed was, again, a composite of those described in BS and T. The salinity of the
current reservoir was chosen to yield a prescribed value of
Δg (relative to the fresh water stream); fractional density
differences Δ up to about 8% were used. Grid parameters S and
f were exploited to achieve the desired values of U' effective

at the interface. The distance of the interface from the grid
was approximately the same for all of our tests, which means
that L' was nominally constant according to correlation
formula (1.2a). Q_c and $\bar{U}_s$ were then adjusted to achieve
equilibrium with the nose of the gravity current located at
the leading edge of the fixed floor. The interfacial flows
were visualised by shadowgraph and typical patterns were
recorded on 35mm film. Measurements were taken of the equilibrium values of Q_c and $\bar{U}_s$ for each of the prescribed values of
Δg, L_c (between about 20cm and 70cm) and U' (for $U'/\bar{U}$ up to
about 20).

2.2. *Interfacial flow patterns*

Photographs of some characteristic patterns appear in
Fig. 4 (a,b,c). The first (a) was typically realized for
turbulent intensities $U'/\bar{U} \lesssim 1/2\%$. It displays all of the
features of currents in non-turbulent streams (see Fig. 1(b)).
Thus the sharp interface at the frontal edge undergoes a
Kelvin-Helmholtz instability, whose engulfment of fluid from
both the stream and the current is responsible for the mixing.
The resulting mixed layer of the thick interface in the downstream flows is apparently stable to shear disturbances.
However, there is turbulence in the stream and this presumably
continues to entrain fluid from the mixed layer. Consequently,
a more complicated pattern must be expected to emerge in the
downstream flows of longer currents. We return to this point
below.

The second (b) is typical of the behaviour found for
$U'/\bar{U} \gtrsim 50\%$. Here we see a picture which is reminiscent of
that presented in T for gravity layers in non-streaming
turbulence: compare Fig. 2(b). The interface is again sharp,
though it now has a wispy-convoluted structure associated with
eddy entrainment and transport of unstirred fluid across the
interface. No Kelvin-Helmholtz waves can be discerned, nor is
there any stable mixed layer in the downstream flows: the
mixing proceeds over the entire length of exposed current.
Notice, however, that in the present experiment the unstirred
fluid is contained within a wedge capped by a sloping interface, rather than a layer capped by a horizontal interface.

The third (c) was obtained with $U'/\bar{U} \simeq 5\%$, an intensity
which is typical of many shear turbulent flows. This picture
displays elements of both (a) and (b). Thus Kelvin-Helmholtz
waves can be identified near the nose whilst turbulent eddies
entrain fluid from the downstream wedge of the current. It
appears as though the turbulence tears the waves apart before
they have evolved sufficiently to overturn and collapse into

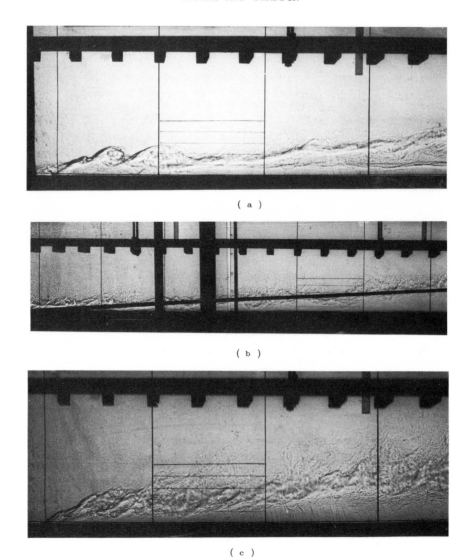

(a)

(b)

(c)

Fig. 4 Characteristic pattern of interfacial mixing as
        revealed by shadowgraph images at different turbulent
        intensities: (a) $U'/\bar{U} \lesssim \frac{1}{2}$% is dominated by shear
        instability at the nose of the current, (b) $U'/\bar{U} \gtrsim 50$%
        is reminiscent of erosive entrainment from gravity
        layers, (c) $U'/\bar{U} \sim 5$% shows a composite picture for
        intensities appropriate to wall-sheared turbulent flows.

a shear-stable mixed layer. In the absence of the thick interfacial layer, the current is exposed to continuing interfacial
erosion throughout its exposed length. Even if the stable
layer was formed - for example, at somewhat smaller turbulent
intensities, it too would be eroded in much the same way.
Indeed, a current of sufficient length might then experience
a series of shear-thickening instabilities, successively
triggered as the stable interface is exposed to erosive
thinning by the stream turbulence. A possible model for these
events and their conjectured evolution is sketched in Fig. 5,
about which we offer comments below.

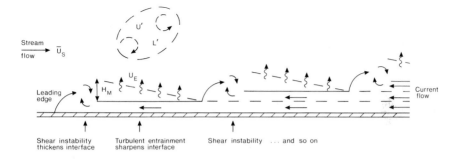

Fig. 5   A possible model of the conjectured regeneration of a
         gravity current travelling in turbulent streams. A
         shear instability at the nose thickens the interface
         which is then sharpened by eddy erosion until another
         shear instability is triggered, etc. Streamwise
         spacing of these events might be expected to scale on
         the shear-mixed layer thickness $H_m \propto \bar{U}_s^2/\Delta g$ and the
         entrainment speed U_E appropriate to Richardson number
         $R_i' = \Delta g L'^2/U'^2$, with adjustments for the gradient
         structure of the mixed layer

Carruthers and Hunt (qv) review various proposals for
turbulent entrainment from stably stratified layers in
connection with their theoretical analysis of the linearised
motions of interfacial and internal waves driven by interfacial
turbulence.

2.3 *Mixing rates*

Fig. 6 presents the dimensionless mixing rate parameter $\Delta g\ Q_c/\bar{U}_s^3$ (shown by BS to be appropriate for non-turbulent currents) plotted against $U'/\bar{U}$. Only for very low intensities $U'/\bar{U} \lesssim 1/2\%$ does the parameter approach a value of 0.2 or so, comparable with that found by BS - equation (1.1). Of course, the physical model conjectured on Fig. 5 would imply that non-turbulent behaviour can only be approached asymptotically. In other words, any turbulence, however weak, inevitably erodes the stable layer and eventually promotes a shear thickening instability. Ignoring Reynolds number effects, the implication would be that, the weaker the turbulence, the greater the interval between the shear instabilities. Thus for a fixed length of current, as in our experiment, it should still be possible to identify a definite lower limit to the turbulent intensity below which only the shear instability at the nose of the current can occur. Further experiments are needed to investigate these ideas.

Turning again to Fig. 6, we see that in high intensity turbulence $U'/\bar{U} \gtrsim 50\%$, the mixing parameter becomes approximately proportional to $(U'/\bar{U})^3$: i.e. slope 3 on this 'log-log' plot. Thus Q_c ceases to depend on $\bar{U}_s$ in this limit, a result which is nicely consistent with our interpretations of the shadowgraph images, namely, that the mixing occurs through erosive entrainment by the eddies in the turbulent stream. Notice also that there is a clear dependence of Q_c on the current length L_c as distinguished by those data for which $L_c \lesssim 45\text{cm}$ and $L_c \gtrsim 45\text{cm}$. The absence of a more refined discriminator betrays our lack of understanding of the physical processes at the time we performed this experiment. With the benefit of hindsight it seems obvious that Q_c should be approximately proportional to L_c in this limit. Supplementary tests are reported below.

A last comment about Fig. 6. Notice that for values of $U'/\bar{U} \simeq 5\%$, a representative order-of-magnitude intensity for turbulence maintained by wall-sheared flows (or by static grids), the mixing rate parameter $\Delta g\ Q_c/\bar{U}_s^3$ is as much as two to four times larger than the non-turbulent value (for the current lengths employed in our experiments). Clearly the modelling of such turbulence effects is crucial if small laboratory replicas are used to estimate the behaviour of large scale industrial and environmental flows.

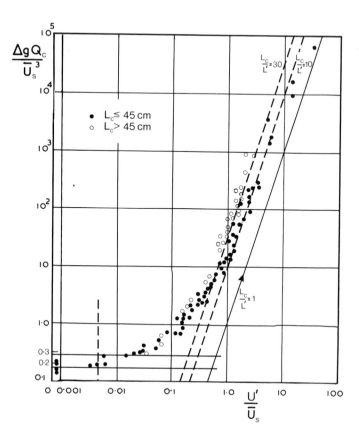

Fig. 6   Dimensionless mixing rate parameters appropriate for
         gravity currents in non-turbulent streams versus
         turbulent intensity. The measurements only approach
         Britter and Simpson's results for $U'/\bar{U}_s \leq \frac{1}{2}$%. The
         mixing flux becomes independent of stream speed for
         $U'/\bar{U}_s \gtrsim 50$%; the lines of slope 3 display dependence on
         current length - see section 4.1(1). For intensities
         appropriate to wall-sheared turbulence $U'/\bar{U}_s \sim 5$%, the
         mixing rate parameter is two-to-four times larger than
         the non-turbulent value.

2.4   Entrainment speeds

   Hereafter we focus attention on the mixing behaviour in
high intensity turbulence, $U'/\bar{U} \gtrsim 50$%, so as to explore the

extent to which eddy entrainment from our gravity current
wedge compares with that from gravity layers. The first step
is to see whether or not Q_c is proportional to L_c, for if it
is (reasonably so), the mixing can be suitably parameterised
in terms of an entrainment speed U_E. Strictly speaking, we
should take the length of the sloping interface rather than its
horizontal projection, but the difference between these is
insignificant compared with other uncertainties in our
measurements. Fig. 7 shows the results of one such test that
we performed and Q_c is indeed proportional to L_c, to a very
good approximation. Notice that there is no indication of any
residual flux when L_c is extrapolated to zero, in accordance
with our earlier findings that shear mixing is absent in this
limit. Perhaps there is also some suggestion in Fig. 7 of a
shortfall in Q_c for $L_c \lesssim 30$cm: if so, this would be consistent
with less entrainment by the decreasing values of U' effective
for transport from shorter wedges - whose interfaces are
further from the oscillating grid.

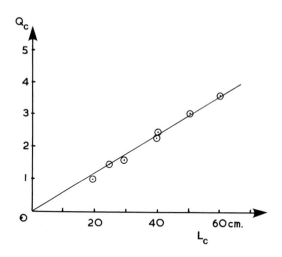

Fig. 7   Mixing flux versus exposed length of current in high
         intensity turbulence $U'/U_s \gtrsim 50\%$. The proportionality
         implies entrainment speed $U_E = Q_c/L_c$ is a suitable
         parameterisation for the mixing (as it is for eddy
         erosion from non-streaming gravity layers).

Having confirmed that U_E is a suitable parameter in this
limit, we repeated a few of the earlier measurements but taking
care, this time, to record the values of L_c. The results
appear on Fig. 8 in the form of a log-log plot of U_E/U' versus
R'_i. The line shown in the figure corresponds to

$$U_E/U' \simeq 2.1 \; R'^{-1.2}_i .$$

Previous data correlations by Turner, Hopfinger and Toly and
McDougall also appear on Fig. 8. The most striking departure
between our findings and the previous ones is not in the value
of the exponent but in the magnitude of the proportionality
coefficient of the supposed power law relation. Our values of
U_E/U' are between two and five times higher than those obtained
in the static tank experiments.

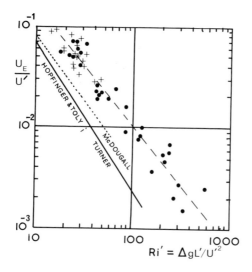

Fig. 8   Entrainment speed measurements ($U'/\bar{U}_s \gtrsim 50\%$) shown as
fractions of the nominal (see text) rms fluctuations
versus turbulent Richardson numbers. Our data for
gravity currents •, and gravity layers +, are shown
here together with a straight-line $U_E/U' \simeq 2.1 \; R^{-1.2}_i$.
Our results are two to five times larger than previous
values measured in static tanks, behaviour which is
attributed to turbulent enhancement due to interference
effects of surface waves

Such a large discrepancy requires explanation. We postulated that it might be attributed to the very shallow submergence
of the oscillating grid in comparison to the previous studies
whose correlation formulae for U' and L' we have adopted thus
far. During the course of our experiments we observed the
motions of the grid excited energetic surface waves.
Conceivably these waves might interfere with the local flows
around the bars of the grid so as to amplify the turbulence
field away from the grid. Since U_E depends on U' raised to
some power between 3 and 4, an enhancement factor of just 3/2
for U' would account for the observed discrepancy in U_E. To
test this hypothesis we performed entrainment experiments on
gravity layers with the oscillating grid located near to the
free surface, as described in the following section.

3. GRAVITY LAYER ENTRAINMENT: SURFACE WAVE INTERFERENCE

3.1 Experimental setup and results

The modifications to our channel are illustrated in the
sketch of Fig. 9. A gravity layer was produced by damming the
channel with end walls which were slotted along their top
edges to accommodate the bars of the oscillating grid. We
found this was necessary to suppress large scale swirling
motions within the test section. Salt solution was continuously
supplied to compensate for the entrainment flux across the
interface. A slow streaming flow was used to maintain the
interfacial buoyancy (without any shear mixing). Measurements
of Q_c were taken under equilibrium conditions achieved for a
range of U' and several values of Δg. The entrainment speed
$U_E = Q_c/L_c$ (where L_c is now the distance between the end walls)
was thus obtained for several different R'_i. Δg is not (quite)
uniform over the streamwise length of the interface. We used
a mean value averaged over the exposed length, deduced from the
bulk flux Q_c and confirmed by local measurements of salinity
obtained with a conductivity probe.

The results are compared on Fig. 8. No systematic differences can be detected between the scattered values of U_E/U'
obtained in the two configurations. To a good approximation
the eddy erosion of current wedges in high intensity turbulence
can indeed be represented by the same entrainment 'laws' as
have been found suitable for gravity layers. Equally, and
intriguingly, it also means that substantial amplifications of
the turbulence are caused by free surface waves interfering
with the local flows around the bars of the oscillating grid.

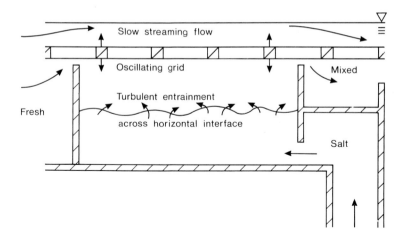

Fig. 9 Modified apparatus for experiments on entrainment from
        gravity layers, to investigate our anomalously large
        entrainment speeds (Fig. 8). The salt water flux
        compensates for eddy transport across the interface and
        the slow stream of fresh water maintains interfacial
        buoyancy without causing any shear mixing.

This finding has since been independently confirmed in a more
detailed study carried out in the static tank described in T.
Enhancement factors of comparable magnitude (two-to-five for
U_E) and even larger, depending on S and f, have been found when
the grid was submerged by less than about 1 mesh size. Some
preliminary results have been reported by Jia Fu and Thomas
(1983).

3.2 Power law entrainment relations

McDougall's (1978) thesis offers an illuminating discourse
about the empirical values which have been attributed to the
exponent $(1 + \alpha)$ in various power law formulae

$$U_E/U' \propto R_i'^{-(1 + \alpha)} \quad .$$

In particular, he argues that his own data for saline layers
(i.e. large P_e') in Turner's tank point to $\alpha \simeq 0.1$, whilst

Turner's data, he says, are better described by $\alpha \simeq 0.3$ rather than $\alpha = 0.5$ as originally suggested by Turner. Accordingly he modestly suggested a compromise, $\alpha \simeq 0.2 \pm 0.1$ to encompass both sets of results.

In a similar spirit, we re-examined Hopfinger and Toly's data and decided that, again, a smaller $\alpha \simeq 0.3 \pm 0.1$ might be more appropriate than $\alpha \simeq 0.5$ which they adopted. This would then imply that α increases from something like 0.1 to 0.2 for the very low $R'_e \lesssim 30$ in Turner's small tank, through 0.2 in our experiments with $R'_e \lesssim 200$, to about 0.3 for $R'_e \lesssim 700$ in Hopfinger and Toly's much larger tank, an apparent trend noted here without any physical explanations which might account for it.

4. DISCUSSION

4.1 Representation of mixing in 'box model' calculations

Practitioners concerned with estimating the dispersion and mixing of large scale gravity driven flows like cooling water discharges and dense gas escapes have, on the whole, compromised the complicated realities and plumped for simple (integral) calculation methods. Most widely used have been the so-called box models; see Webber (1983) for a recent review. These approaches postulate shape parameters for the geometry of the interface, typically rectangular in side-projection. Exchange fluxes across the interface are prescribed in terms of 'lumped-parameter' closures such as entrainment relations. Although their physical basis may be naïve, the resulting description in terms of ordinary differential equations yields semi-analytical formulae suited to rapid evaluation of sensitivity trends. Convenience, then, has so far given box models an edge over apparently more refined schemes couched in terms of partial differential transport equations and local turbulence closures, e.g. eddy diffusivity models like "k - ε". However, there does seem to be an increasing interest in these latter methods, because in principle, they avoid the need for dubious presumptions about the interfacial geometry. Whether or not the latter can be accurately resolved and reliably computed by such schemes is another matter!

To judge from Webber's review, it seems there is still considerable scope for improvements in the physical plaus-ibility of box model representations of the interfacial fluxes. In particular, a clear distinction has not usually been drawn between entrainment of external fluid into the current and

detrainment of current fluid into the surroundings. Thus
turbulence in the surroundings has been supposed to provide a
suitable measure for parameterising the dilution of the
gravity current. Now this is clearly at odds with our experimental findings that external turbulence erodes the stable
buoyancy interface and detrains fluid from the current but
without causing dilution of the current fluid. Internal
turbulence, on the other hand, gives rise to entrainment of
surrounding fluid into the current and so provides for such
dilution. The flow in the current fluid was non-turbulent in
our bench top experiments, whereas substantial intensities are
sustained in the large scale flows of practical interest.
Typical sources of turbulence arise from disturbances introduced in the establishment zone of the gravity driven flow
(e.g. jetting or collapsing motions), convective forcing by
heat transfer at the boundary (e.g. cold gas spreading over
warm ground, or cold wind blowing over a warm water plume) and
boundary friction (e.g. ground stress on dense gases or wind
stress on warm water plumes). For the last of these, at least,
suitable measures of the internal turbulence are given by the
friction speed and the depth of the current.

In general, then, both an entrainment flux and a detrainment
flux must be expected and these fluxes need to be explicitly
parameterised in terms of the internal and external turbulence
quantities respectively. Although not in a position to write
down specific proposals here (i.e. without more information
about the events postulated on Fig. 5), we briefly outline some
of the scaling implications which have been brought to light
by our experimental studies.

(1) Mixing at high turbulence Richardson numbers

Fig. 10 shows the definition sketch for a simple idealisation of the gravity current wedge which we observed in high
intensity turbulence. For the conditions of our experiments
the current can be assumed non-turbulent so its density is
conserved and $\rho_c = \hat{\rho}_c$; hats denote reference or source conditions. However, detrainment of current fluid gives rise to
varying stream density $\rho_s(x)$, where x is horizontal coordinate
measured from the nose of the current. This in turn leads to
varying interfacial buoyancy which we choose to represent in
terms of Boussinesq reduced gravity parameters, as follows:

$$\left.\begin{aligned}
\Delta g_s(x) &= g(\rho_s(x) - \hat{\rho}_s)/\hat{\rho}_s \\
\Delta g_c &= \Delta \hat{g}_c = g(\rho_c - \hat{\rho}_s)/\hat{\rho}_s
\end{aligned}\right\} \tag{4.1}$$

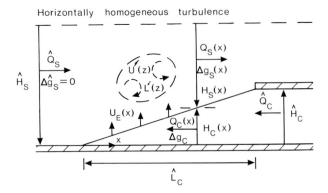

Fig. 10 Definition sketch for simple idealised modelling
analysis of the gravity current wedge observed in
high intensity turbulence

Integral relations for flux conservation at any station x
can also be written down immediately, viz:

buoyancy

$$\Delta g_S(x) Q_S(x) = \Delta \hat{g}_C \, Q_C(x) \quad , \qquad (4.2)$$

stream volume

$$Q_S(x) = \hat{Q}_S + Q_C(x) \quad , \qquad (4.3)$$

current volume

$$Q_C(x) = \int_0^{} U_E(x) \, dx \quad ; \qquad (4.4)$$

see Fig. 10. Notice that shear mixing near the nose of the
current has been omitted in this limit of high intensity
turbulence. Because the source flux is totally entrained
across the interface, we also have

$$\hat{Q}_C = < U_E > L_C \qquad (4.5)$$

where the carot brackets denote an effective value averaged
over the length of the current.

Streamwise variations in all flow quantities can be sensibly linearized in the limit of small fractional depths $\hat{H}_c/\hat{H}_s$ and fluxes $\hat{Q}_c/\hat{Q}_s$; this approximation is not valid near the nose of the current, a region which has been excluded by hypothesis in this analysis. With such a restriction, it is possible to deduce salient features of the entrainment scalings without explicitly invoking conservation of momentum. An empirical closure is still required for the entrainment speed and here we adopt the power law relation suitable for high R'_i. In terms of the effective quantities, then

$$< U_E >/< U' > = C_E < R'_i >^{-(1 + \alpha)} \qquad (4.6)$$

with

$$< R'_i > = (1 - < \Delta g_s(x) >/\Delta \hat{g}_c) \; \Delta \hat{g}_c < L'>/< U' >^2$$

and

$$< \Delta g_s(x) >/\Delta \hat{g}_c = \tfrac{1}{2} \hat{Q}_c/\hat{Q}_s$$

for linearised currents. Combining equations (4.5) and (4.6), the mixing flux achieved in non-sheared turbulent flows is given by

$$\Delta \hat{g}_c \hat{Q}_c/ < U' >^3 = C_E(1 + \tfrac{1}{2} \hat{Q}_c/\hat{Q}_s) < R'_i >^{-\alpha} L_c/ < L' >$$
$$(4.7)$$

when scaled in the appropriate units. The fractional flux term, assumed small in this argument, appears here because of the streamwise variations in interfacial buoyancy.

Compare this expression with BS finding (formula 1.1) for non-turbulent sheared flows

$$\Delta \hat{g}_c \; \hat{Q}_c/\bar{U}_s^3 \simeq C_s \qquad (4.8)$$

where $\hat{Q}_c$ now relates to the mixing flux which is confined to the nose zone of such flows. Recall that the relation (4.8) can be taken to imply that a fixed fraction of the available kinetic energy is converted to potential energy of buoyant mixing. Equation (4.7) can be interpreted in the same way, but only when parameter α is identically zero (Turner, 1979).

The non-zero α found in practice (about 1/5 according to our results) has been interpreted (Linden, 1973) as representing the fraction of kinetic energy which is unavailable for mixing because it goes to recoiling motions of the interface and rebounding motions of eddies following their impingement on the interface. Linden put forward scaling arguments to suggest $\alpha = 1/2$ in agreement with Turner's evaluation of the experimental results for large Peclet number turbulence. However, we suspect that any simple power law would be inappropriate asymptotically in this limit. After all, the limit is also realisable in practice for layers approaching immiscibility. With this interpretation, it would seem more plausible to represent the interfacial exchanges in terms of a turbulent entrainment flux which is opposed by a buoyant detrainment flux - of discrete drops in the immiscible limit. Experimental results obtained with stirred kerosene and water layers have been modelled in this fashion by Thomas and Kerr (to be published). Indeed the same comments also apply to the shear-mixing flux achieved in the immiscibility limit and a similar modelling approach was put forward in Thomas (1982) for the flux of bubbles transported by a free shear layer on the edge of a submerged jet.

The ratio of mixing fluxes, turbulent-to-shear, as given by relations (4.7) and (4.8) is

$$\frac{C_E}{C_s} \frac{L_c}{< L' >} < \frac{U'}{\bar{U}_s} >^3 < R'_i >^{-\alpha} \tag{4.9}$$

when the fractional flux term $1/2 \; \hat{Q}_c/\hat{Q}_s$, assumed small here, is neglected. Clearly it is the cube root of this expression which should be used to correlate experimental data, instead of just $< U'/\bar{U} >$ as appears in the abscissa of Fig. 6. (We cannot explicitly test the validity of the scaling relation (4.9) because of the inadequacies in the present dataset - recall our comments in section 3). However, the practical import of this ratio should be obvious. L_c increases with distance from the source, whilst $\bar{U}_s$, effectively the relative spreading speed of the gravity current, decreases because of the dispersion and the turbulent detrainment - both of which make the cloud thinner and hence reduce $\bar{U}_s$ which scales on $(\Delta g_c \; H_c)^{\frac{1}{2}}$; see BS. Notice that internal turbulence, and hence entrainment of external fluid, also declines with decreasing

$\bar{U}_s$. The implication is that far field mixing is dominated by
interfacial detrainment when external turbulence is maintained
at high Richardson number.

Gravity driven flows travelling over the boundary responsible for the external turbulence (e.g. dense gas released into
wind) need to be distinguished from those which travel over
the remote boundary (e.g. warm water plume on the surface of
shallow water flow). In the dense gas problem a more appropriate model would allow for decay of the turbulence as the
boundary stress in the wind relaxes when it flows over the top
of the current. In this case, the interfacial Richardson
number increases with streamwise distance from the nose, so R_i'
can be assumed to remain high if initially so. An outline
consideration of the effects of decay on mixing is given below
in section 4.2.

Although mixing at low R_i' was not explored in our apparatus,
a few comments are offered below on such a generalisation of
the modelling.

(2) Mixing at low turbulence Richardson number

Instead of the power law relation (4.6), we can adopt an
expression favoured by the 'box' modellers, namely,

$$U_E/U' = C_{E1}(1 + C_{E2}\, R_i')^{-(1 + \alpha)} ; \qquad (4.10)$$

see Webber's review. With $C_{E1} \simeq 0.8$ appropriate for neutrally
buoyant conditions (Turner, 1979), and supposing $\alpha \simeq 1/5$ as
before, then $C_{E2} \simeq 3/2$ recovers the entrainment speeds measured
in static tanks at high R_i'. In principle, then, such an
interpolation formula encompasses interfacial entrainment at
any R_i'. Here, however, we shall only inspect the scaling
implications for very small values, when

$$< U_E >/< U' > \simeq C_{E1} , \qquad (4.11)$$

i.e. essentially neutral interfaces. Admittedly, the box model
representation becomes suspect in this limit because turbulent
eddies no longer erode and sharpen the interface, but instead

diffuse and thicken it. Putting such reservations to one side,
we find the ratio of mixing fluxes sustained by turbulence and
by shear (cf. equation 4.9) is now given by

$$\frac{C_{E1}}{C_s} \left(\frac{\Delta \hat{g}_c L_c}{\bar{U}_s^2} \right) < U'/\bar{U}_s > . \tag{4.12}$$

This expression appears to imply that each of three ways in
which R_i' can be rendered small (i.e. large U', small $\Delta \hat{g}_c$
or small L') has a different consequence for the mixing ratio.
Thus large U' implies that turbulent mixing ultimately dominates, whereas small $\Delta \hat{g}_c$ seems to imply shear-mixing does and
L' appears to have no effect. The first accords with physical
expectations. The second inference is flawed because allow-
ance is needed for the dependence of $\bar{U}_s$ on $\Delta \hat{g}_c$ and (H_c), i.e.
that $\bar{U}_s \sim (\Delta g_c H_c)^{\frac{1}{2}}$ as noted earlier. If we now suppose that
Δg_c becomes small with H_c remaining fixed, then turbulent
mixing indeed dominates, as expected. The third implication
presumably reflects on the physical limitations of the entrainment hypothesis: a reasonable additional constraint would be
that L' be small compared with H_c for diffusive turbulent
mixing to occur.

Considerable practical significance attaches to knowing more
precisely the conditions under which external turbulence
provides for diffusive thickening of the current instead of the
erosive thinning which we observed. Thus, in his model based
on field observations of dense gas dispersion and mixing, van
Ulden (1974) proposed that an initial phase of buoyancy
dominated spreading gives way to quasi-neutral diffusive thickening sustained by atmospheric turbulence and thence to declin-
ing levels of boundary concentration. Britter's (1979)
laboratory experiments supported this picture and
Colenbrander's (1980) computational scheme also presumed
diffusive eddy transport with buoyancy damping across a thickening interface.

One indication as to which behaviour should be expected can
be had from a simple consideration of the energetics, as
proposed by Crapper and Linden (1974). They suggested the
transition be identified with maximal extraction of buoyant
potential energy flux per unit input of turbulent kinetic
energy flux. Now this ratio is proportional to $R_i' U_E/U'$ and

for entrainment relation (4.10), its maximum value occurs at $R_i' = (\alpha \, C_{E2})^{-1}$. Admittedly, there are reservations about the sensitivity to uncertain parameter α (i.e. when α goes to zero, there is no transition). However, the values quoted earlier ($\alpha \simeq \frac{1}{5}$, $C_{E2} \simeq \frac{3}{2}$) would suggest diffusive thickening might occur when R_i' is less than about 3 and erosive thinning otherwise. Crapper and Linden's own experimental results suggested a value of about ten.

Gradient transport calculations near to this transition condition have been reported by Puttock (1976).

4.2 Effects on the mixing of decay in the external turbulence

Decaying turbulence in the external stream is more appropriate for gravity flows spreading over the turbulence generating boundary. A suitable laboratory experiment could be had by using a static grid to intercept all of the streaming flow upstream of the nose of the gravity current. With this idealisation in mind consider the empirical relations

$$U'/\hat{U}' = (1 + x/x_*)^{-m} \; , \; L'/\hat{L}' = (1 + x/x_*)^{-n} \qquad (4.13)$$

to represent the decline in rms fluctuations and the growth in eddy size with streamwise distance. Note that x_* defines both a virtual origin for the turbulence and a measure of the 'turnover' distance or Lagrangian length scale. Elementary parameterisations of the turbulence kinetic energy budget provide for a physical constraint that $m + n = 1$, i.e. that the turnover time scale of the eddies is proportional to their transit time. For typical values of m and n we quote Comte-Bellot and Corrsin's (1966) experimental results which are consistent with $m \simeq 5/7$, $n \simeq 2/7$; earlier data suggested $m \simeq n \simeq \frac{1}{2}$.

Representing the mixing behaviour simply in terms of decay relations (4.13) taken in conjunction with the conservation integrals (4.2 to 4.4) and an entrainment relation (4.10) is extremely primitive. It neglects any turbulence generated by interfacial friction, perhaps further complicated by a sequence of shear-mixing events (as suggested earlier). It also overlooks the fact that the buoyancy flux is expressed within a growing buoyancy boundary layer capping the interface; see Piat and Hopfinger (1981). Finally, it fails to take account of the 'blockage' boundary layer measured in the moving

wall experiments of Thomas and Hancock (1977), in which the
normal component fluctuations are inhibited by the boundary, as
analysed by Hunt and Graham (1978).

If these shortcomings are accepted, the previous estimates
can be readily extended, by elementary analysis if $\alpha \ll 1$ or
$(1/2 - \alpha) \ll 1$ - a reasonable approximation here. The details
will not be reproduced; we merely note that the solution can
be expressed in terms of elementary functions for a family of
paired values of m and n, each with sum unity. The essential
physical point is that attenuation of the mixing flux due to
turbulence decay cannot be reasonably neglected unless the
current length is short compared with the turnover distance,
i.e. $\hat{L}_c/x_* \ll 1$. In terms of the parameters noted earlier,
this condition is

$$(\hat{L}_c/\hat{L}') \; (\hat{U}'/\hat{\bar{U}}_s) \ll 1 \qquad\qquad (4.14)$$

irrespective of the value of R'_i. Compare the ratio of
turbulent-to-shear mixing fluxes at low R'_i, expression (4.12),
which is proportional to

$$(\hat{L}_c/\hat{H}_c) \; < \hat{U}'/\bar{U}_s > . \qquad\qquad (4.15)$$

The implication would seem to be that when turbulent mixing
becomes significant, decay cannot reasonably be neglected
unless $\hat{L}'/\hat{H}_c \gg 1$. However, in this case, the entrainment
hypothesis itself becomes implausible, as noted earlier.

Suppose we take 5% to be a typical value for the turbulence
intensity of an approaching shear flow and $\hat{L}'$ as some fraction
of $\hat{H}_c$, corresponding to the energetic eddy size at the height
of the interface. Self-consistency thus demands that decay
effects must be incorporated in simple box models. The calculations also indicate that the attenuation at large $\hat{L}_c/x_*$ is
proportional to $\hat{L}_c/x_*$. This means a gravity current whose
planform dimension is, say, ten turnover distances experiences
about one-tenth of the total detrainment flux achieved in the
absence of decay. (The order of magnitude of this estimate,
deduced with unreal but analytically convenient m = 1/3,
n = 2/3, is insensitive to the parameter α).

4.3 Implications for mixing of three-dimensional gravity currents

   A three-dimensional gravity current from a localised source in a turbulent stream is sketched on Fig. 11. Here the current is travelling along the free boundary remote from that generating the turbulence, as appropriate for warm water discharged into a cold stream and in the absence of wind stress at the free surface. For large scale discharges, such as heated water from power stations, the near field flows close to the source are highly agitated and the contaminant is locally mixed throughout the depth of flow (Ewing, 1982; Rodgers, 1983: private communications). Further away, however, the buoyancy of the discharged fluid is often sufficient to promote and sustain a surface plume layer as shown in the sketch. At large distances from the source, the steamwise component velocity of the plume approaches that of the streaming flow: i.e. $U_c(x) \to U_s = U$, say, can be taken to define the far-field flow.

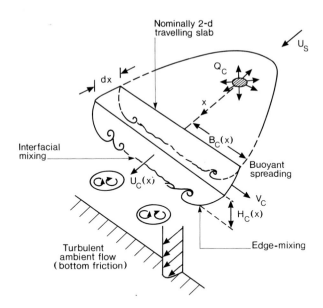

Fig. 11   Definition sketch for three-dimensional gravity
          current from a localised source in streaming
          turbulent flow

An outline argument below suggests that the far field mixing
must usually be dominated by interfacial turbulent detrainment
from the current into the streaming flow. Our interpretation
accords with recent field measurements described by Pickles
and Rodgers (qv) who have also developed an eddy diffusivity
model of their observations. However, these findings are in
striking contrast to some earlier calculations which predicted
the asymptotic mixing to be described by entrainment, not
detrainment. Quantitative modelling estimates (to be
published) confirm the present conclusions and also incorporate
additional effects of surface jetting and plume turbulence.

The far field spreading behaviour can be addressed within
the framework   of   two-dimensional gravity currents,
following, for example, Ewing (1981: private communication).
Consider a slab of current fluid, half-width $B_c(x)$, depth
$H_c(x)$, streamwise incremental length dx and travelling with the
uniform speed U of the flow. In this reference frame the side-
ways spreading is locally two-dimensional and the spreading
speed $V_c(x)$ can be inferred from a tangency condition applied
along the edge between the current and stream flows, viz

$$U \frac{dB_c}{dx} = V_c = k(\Delta g_c H_c)^{\frac{1}{2}} \qquad (4.16)$$

where $k \sim 1$, as in BS. In reality the edge is a line of
surface flow convergence where both the current fluid and the
stream fluid are drawn downwards into the shear instability
shown in the sketch. As to how this sink-like character of the
edge affects the validity of the assumed tangency condition is
not discussed here.

To continue the argument, multiply both sides of equation
(4.16) by $(UB_c)^{\frac{1}{2}}$ and integrate to give

$$U B_c(x) = \left\{ \frac{3}{2\sqrt{2}} k \int_0^x (\Delta g_c Q_c)^{\frac{1}{2}} dx \right\}^{2/3} \qquad (4.17)$$

where $Q_c = 2UB_c H_c$ is the plume flux and x is assumed measured
from a virtual origin where $B_c(0) = 0$. Assume that the current
buoyancy flux $\Delta g_c Q_c$ is conserved, although strictly this
condition is appropriate only for entraining currents. Then

$$U\,B_c(x) = \left(\frac{3}{2\sqrt{2}}\,k\right)^{2/3} (\Delta g_c Q_c)^{1/3}\,x^{2/3}\,,$$

so B_c increases as $x^{2/3}$ and hence V_c decreases as $x^{-1/3}$. To estimate $H_c(x)$, suppose that Q_c is conserved (approximately), whence H_c decreases as $x^{-2/3}$. Given these estimates, what are the implications for the mixing behaviour?

Firstly, the shear-mixing at each of the edges sustains a flux per unit length of current which scales as $V_c^3/\Delta g_c$, in line with earlier deductions by Britter (1979) and by Ewing (loc cit). Notice that this contribution varies as x^{-1} according to the estimates deduced above. Secondly, the external turbulence of the stream detrains current fluid, provided that the inter-facial Richardson number $R_i' \gtrsim 3$, or so; for $R_i' \lesssim 3$, a diffusive flux across the interface is expected. The detrainment flux at high R_i' can be estimated when B_c is identified with L_c. If variations in U', L' and interfacial buoyancy are neglected, R_i' is independent of streamwise position and hence U_E is also. In this case the flux per unit length due to interfacial turbulent mixing is proportional to B_c, and this increases as $x^{2/3}$. The shear mixing contribution, remember, falls off as x^{-1} according to the above estimate.

Although the approximations may be crude, the implication is clear: the far field interfacial exchange flux of buoyant surface plumes in streaming turbulent flow is dominated by eddy erosive entrainment of current fluid, provided only that the interfacial turbulence Richardson number exceeds three or so and that wind-induced turbulence in the current is small compared with that in the streaming flow, as may often be the case.

ACKNOWLEDGEMENTS

Our appreciation to David Carruthers, Jim Rottman and Julian Hunt of DAMTP, University of Cambridge for their constructive criticisms. Also to David Ewing, Ian Rodgers and Jeremy

Pickles of Central Electricity Laboratories for numerous
discussions about their field data.

REFERENCES

Benjamin, T.B.B. (1968) *J. Fluid Mech.*, **31**, 209.

Britter, R.E. (1979) *Atmos. Environment,* **14**, 779.

Britter, R.E. and Simpson, J.E. (1978) *J. Fluid Mech.*, **88**, 223.

Colenbrander, G.W. (1980) 3rd Internat. Sympos. Loss Prevention
    (Basle, September).

Coles, D. and Barker, S. (1974) *Bull. American Physical Soc.,*
    19 (10), 1146.

Comte-Bellot, G. and Corrsin, S. (1966) *J. Fluid Mech.*, **25**,
    667.

Crapper, P.F. and Linden, P.F. (1974) *J. Fluid Mech.*, **65**, 45·

Hopfinger, E.J. and Toly, J.A. (1976) *J. Fluid Mech.*, **78**, 155.

Hunt, J.C.R. and Graham, J.M.R. (1978) *J. Fluid Mech.*, **84**, 209.

Jia Fu and Thomas, N.H. (1983) Proc. Second Asian Conf. on
    Fluid Mechanics, (China, October), 906.

Linden, P.F. (1973) *J. Fluid Mech.*, **60**, 467·

Linden, P.F. (1979) *Geophys. Astrophys. Fluid Dynamics,* **13**, 3.

McDougall, T.J. (1978) Ph.D. dissertation, Univ. of Cambridge.

McDougall, T.J. (1979) *J. Fluid Mech.*, **94**, 409.

Piat, J-F. and Hopfinger, E.J. (1981) *J. Fluid Mech.*, **113**, 411.

Puttock, J.S. (1976) Ph.D. dissertation, University of
    Cambridge.

Thomas, N.H. (1982) Proc. Internat. Conf. Hydraulic Modelling
    (Coventry, September). Paper E5, BHRA.

Thomas, N.H. and Hancock, P.E. (1977) *J. Fluid Mech.*, **82**, 481.

Turner, J.S. (1968) *J. Fluid Mech.*, **33**, 639.

Turner, J.S. (1979) 'Buoyancy Effects in Fluids'. Second Edition, CUP.

van Ulden, A.P. (1974) 1st. Internat. Sympos. Loss Prevention (The Hague, May), 221.

Webber, D.M. (1983) SRD Report, R243.

A VIVID MECHANICAL PICTURE OF TURBULENCE

R.S. Scorer

(Imperial College, London)

1. THE BASIC CONCEPT

Turbulence must be vorticity because it consists of motion whose energy is locally resident in the fluid and it stays there, usually, and degenerates into heat. It cannot be irrotational because irrotational motion can be brought absolutely to rest by bringing the boundaries to rest.

In an inviscid fluid vortex lines and tubes move with the fluid, except when density gradients are present and we shall discuss those later. So turbulence moves with the fluid, and it is inappropriate to refer to fluctuations of velocity representing transmitted waves, such as sound waves, as turbulence. They do not cause diffusion. Equally, fluctuations caused by the motion past an irregular boundary are irrotational, and should not be regarded as part of the turbulence.

All vorticity is obviously not turbulence, which is the part of the motion which we can't comprehend in detail or which we consider it would serve no purpose to specify in detail. This attitude is summed up by describing turbulence as chaotic vorticity, a very subjective term, with the additional qualification that it causes diffusion: it causes any material or attribute such as heat carried by the fluid to be spread on a continuously finer scale forever.

Of course it doesn't go on forever because soon viscosity takes over and turns the kinetic energy into heat, and the material and heat are subject to molecular diffusion.

2. THE ENERGY CASCADE

It has been suggested that a state of turbulent equilibrium
would exist if viscosity did not destroy the motion at the
small scale end of the eddy size spectrum; but since it does
so the equilibrium is restored by a cascade of energy from the
large eddies which, on the same time scale, are not affected
by viscosity. That is a fallacy. The cascade is an essential
feature of the turbulence itself and if there were no viscosity
the energy would continue to pass to the small eddies; the
consequence of that would be what is picturesquely called the
ultra-violet catastrophe. If the world had no means of
storing heat in molecular motion the small eddies would become
ultra-violent indeed.

A vortex tube representing a small eddy extracts kinetic
energy from its surroundings if it is stretched in the course
of being carried by the fluid. If it is shortened it will
feed energy to the surrounding motion and increase the vorticity of some larger scale motion - either an eddy or possibly
the mean shear - and lose some of its K.E. of rotation in the
process. If the turbulence were truly isotropic at $t = 0$,
for every tube being stretched there would be another being
allowed to contract so that just as much energy would be fed
up the size scale as was fed down it. There would be no
Reynolds' stresses, for in perfect isotropy every positive u'
would be as often associated with -w' as with +w'.

But such a situation cannot last. Thus if at $t = 0$ there
are two tubes of vorticity OA and OB in a flow in which
(Fig. 1) O is at rest and the particles on AB are moving in
the direction AB, at a later time the tubes will be at OA'
and OB' and both will be in a state of being stretched.

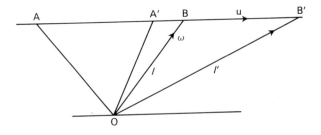

Fig. 1

The tube that is stretched from OB to OB' has been elongated from length ℓ to ℓ' and laterally contracted from radius r to r' so that if the fluid is incompressible

$$\ell r^2 = \ell' r'^2$$

The conservation of angular momentum requires that the initial and final angular velocities and kinetic energies are related by

$$r^2\omega = r'^2\omega'$$

and

$$T = \ell r^2 \omega^2, \qquad T' = \ell' r'^2 \omega'^2$$

$$= \frac{\ell'^2}{\ell^2} T$$

so that it now possesses more energy than before. It is also smaller than it was because r has been reduced to r' and although the length of OB has increased the tube is infinitely long and tortuous, and the size of the tortuosities, which are also decreasing in length scale, is the relevant other measure of the size.

The tension in the tube is greater because the pressure across the ends at O and B has been decreased as the interior pressure is decreased by the intensification of the rotation. Work has to be done to stretch every bit of the tube, and this is done by forces which act so as to reduce the velocity of B away from O, which means decreasing the vorticity of the larger eddy, or mean motion, which is carrying the fluid along AB in the first place.

It is worth noting in passing that the so-called return to isotropy is a misnomer: it is a return to equipartition of energy as between rotation around the three coordinate directions.

Thus the turbulence quickly becomes non-isotropic at every size scale as a result of the vorticity of the larger scale motion. Energy is not only passed to the smaller eddies but the size of each eddy is also decreased.

3. REYNOLDS STRESSES

What has been said is strictly true of every infinitesimal tube of fluid surrounding every vortex line, and this is not a model in the sense that it would contain one or more hypothesis. It is a picture which is correct, quantitatively; but it cannot be made quantitative in practice because the situation is too complex. We do know, however that Reynolds stresses do appear, and their strength gives us some idea of the importance of the mechanisms described.

The correlations set out in Reynolds' equations for the mean motion are evident in the behaviour of the vortex tubes of the eddies. Thus if PQ is a vortex tube being stretched by a shear flow, the flow due to the tube on the far side of it has a negative correlation between the fluctuations u' and w' because, as drawn, u' is negative. On the near side u' is positive and w' is negative. A vortex tube like RS exerts no Reynolds stress, nor does one like PR, because neither is being stretched (although RS soon will be).

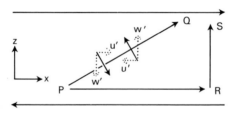

Fig. 2

The tubes are like bits of elastic, because if the parcels at their ends which are being drawn apart by the surroundings were allowed to move back towards each other the energy would be returned to the mean motion. Actually the tubes are bound into the surrounding fluid by knots and tortuous bundles of other spaghetti-like vortex tubes if the turbulence is worth its salt, and so if the surrounding motion were such that a parcel of fluid were subjected to a distortion which was the reverse of the one it had just experienced, it would not experience further Reynolds stresses but would receive its energy back again. The fibres, or elastic filaments of the turbulence, which are the chaotic vortex tubes, therefore provide a visco-elastic system of stresses. This would be more

evident the larger the eddies causing the Reynolds stresses:
the smaller these eddies are the sooner is their orderly
stretching dragged out of shape by the surrounding chaos, and
the sooner is the orderly elastic behaviour   destroyed.
There is no analogy with viscous mechanisms here, and "mixing
length" theories or "displaced parcel" theories cannot be
made to prove that turbulence is visco-elastic, however
cleverly they are manipulated to make it possible. Of course,
again, the vortex tube spaghetti theory can't give a quantitative estimate of its magnitude but it can confidently state
that the visco-elasticity is a reality, and its presence is
probably worth demonstrating and if possible quantifying.

4.   ENTER STABLE STRATIFICATION

The presence of a stable stratification makes it possible
for additional vorticity to be created by gravity. The
vorticity equation for an incompressible inviscid fluid is

$$\frac{D\omega}{Dt} = (\omega.\underline{\mathrm{grad}})v + \frac{1}{\rho}\ \underline{\mathrm{grad}}\ \rho \times (\underline{g} - \underline{f})$$

where $\underline{f}$ is the fluid acceleration, and term in $\underline{f}$ is more
usually written in terms of $\underline{\mathrm{grad}}\ p$ by means of the equation of
motion. Here we put it together with gravity so that the
mechanism is obvious, for it merely serves to create additional
vorticity in exactly the same way as gravity does.

The consequence of this is that if we start off with a fluid
which is in horizontal motion with the flow in surfaces of
constant density, and the vorticity in the same surface (i.e.
horizontal velocity and vorticity), the equation shows that
according to the first term the vortex lines move with the
fluid and according to the second new vorticity is created
only perpendicular to $\underline{\mathrm{grad}}\ \rho$, i.e. in the surfaces of constant
density. The vortex lines therefore always remain in surfaces
of constant density.

This will not be true if grid turbulence is introduced
because it is caused by separation of the boundary layers into
the fluid, and that introduces new vorticity. Nor will it be
true if the stratification is destroyed in such a way that
the surfaces containing the vortex lines are ruptured so that
discontinuities of density are created.

If however we can imagine the motion to remain laminar, but
be chaotic and therefore turbulent, the vortex lines will be

contorted and knotted up in exactly the same way as the
isopycnic surfaces.

That is really taking the model too far, but it does raise
interesting possibilities in the atmosphere when turbulence
is caused by overturning billows.

The most important effect of density stratification is to
introduce the possibility of wave motion, for an eddy may
oscillate about static equilibrium and transmit its kinetic
energy away from the fluid containing it. On the whole this
will not alter the eddy energy density in a large region
because it will be transmitted more or less equally in all
directions. But it does have consequences for the diffusion
of material and properties carried by the fluid.

5.   PENETRATIVE MOTION

In order to transfer material, eddies must penetrate their
environment. We may think of what vorticity is required to
transfer parcels of fluid upwards and downwards through their
environment. The fluid rising through its surroundings is
surrounded (Fig. 3) by horizontal vortex rings. In the case
of a blob of rising lighter fluid these have been generated by
the density gradients in the presence of gravity (Fig. 4)
according to the term $\frac{1}{\rho} \underline{\text{grad}}\ \rho \times \underline{g}$ in the vorticity equation,
which is rotation around the line where the isopycnic inter-
sects the horizontal.

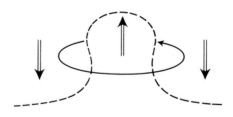

Fig. 3

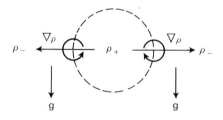

Fig. 4

If we use a grid of vertical rods only to generate turbulence, we shall have to wait until the tendency towards
equipartition produces some horizontal chaotic vorticity, for
only then will material be diffused vertically.

Reversing this argument we note that if equipartitioned
turbulence is created by a grid in a stably stratified fluid
some of the energy of rotation about horizontal axes will be
oscillatory, but this will not be true of rotation about the
vertical. The consequence will be that horizontal diffusive
transfer will be slightly reduced while vertical transfer will
be much more inhibited. If there are any absorbent boundaries
or limits to the region of turbulence in a more extensive
stratified fluid, turbulent energy which can assume wave form
can be transmitted out of the region but it can't perform
diffusion of any kind if it is transmitted into a formerly
non-turbulent region because the vorticity is only oscillatory,
except possibly for the occasional "splash" where additive
effects of waves crossing each other produce overturning and
penetration.

The most important effect of stable stratification on grid
turbulence will be a significant reduction in vertical
diffusion. The reduction in turbulent intensity will be much
less easy to detect, as also will be the reduction in horizontal diffusion.

6. BUOYANT CONVECTION

There is a converse to this last result which states that
if the turbulence is thermal convection, all the turbulent
energy generated by the buoyancy forces will initially take
the form of horizontal vortex lines, i.e. perpendicular to g.
This means that vertical diffusion will be very much more
effective than horizontal diffusion, which, it may be said,

is obvious. What is less obvious is that such turbulence will
not initially transfer any momentum vertically because
horizontal vortex lines do not produce   Reynolds stresses in
horizontal shear flow.

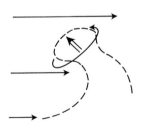

Fig. 5

    However, once the material penetration has begun the
velocity of the penetrating masses relative to the surroundings
into which they have risen acquires a horizontal component
when they rise through shear. They therefore clothe themselves
in an additional collection of vortex lines which consist of
vortex rings whose axes are in the direction of the shear
(Fig. 5). The position of the added rings of vorticity is
depicted in Fig. 5, but this is drawn so as to give special
emphasis to the rings in vertical planes. What actually
happens is that the rising towers of fluid penetrating upwards
are accelerated down the direction of the shear and so,
relative to their starting point they lean over down the shear.
A cumulus tower rising through a wind increasing with height
leans downwind, but it is travelling upwind relative to its
environment at every level (Fig. 6).

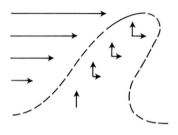

Fig. 6

Figuratively one might describe this penetrating tower as being dragged downwind, and with this in mind some people have attributed to it a drag coefficient as if it possessed a turbulent wake. But a mass of fluid penetrating through similar fluid has no wake. It mixes furiously with the fluid into which it is advancing and is retarded by dilution. It is a puff, which is very like a thermal, a rising mass of buoyant fluid, in velocity distrbution, and this makes it possible to analyse the horizontal and vertical components of motion together. The consequence is as expected, for a puff dies out (loses its velocity) much more rapidly than a buoyant thermal whose upward momentum is continually renewed. Much of the vortex ring energy around a penetrating mass is partitioned and dissipated by the turbulence of the mixing with the surroundings, but gravity is continuously replenishing the horizontal turbulence vortex tubes, and they achieve quite a bit of vertical transfer before they are partitioned and subsequently dissipated: only after some penetration through the shear has been achieved are rings of vorticity representing the transfer of momentum generated. Therefore the vertical transfer of buoyancy and material is much more effective than the vertical transfer of horizontal momentum. Or to use the conventional terminology of K-theory, $K_M/K_H \ll 1$ in thermal convection. In a stratified fluid, as we have already seen $K_M/K_H \ll 1$, and as we know that stable waves can actually transfer horizontal momentum vertically when the phase of the wave is displaced horizontally at higher levels, we see no reason why K_M/K_H should not be two or three orders of magnitude away from unity in a powerfully stable stratification or a fiercely unstable one.

7. BILLOW MECHANICS

It is not out of place to include relevant thoughts on billows, which are unstable gravity (K-H) waves in a vortex layer. It is often asserted that shear in itself is a cause of dynamic instability. This is not so, for uniform shear is not unstable, and is positively stable if there is a stable density gradient present. The cause of instability in a vortex layer, or sheet, is that, assuming it to be horizontal in the undisturbed state, the crests and troughs of wave disturbances are propelled in opposite directions by the vorticity itself (Fig. 7). Thus the fluid containing the vorticity is accumulated at the downward sloping nodes and is drawn away from the upward sloping ones, and the layer becomes a row of vortices, which is itself unstable to disturbances of much longer wavelength through the same mechanism. Indeed all wavelengths are unstable in the first place except those that are short compared with the vortex layer thickness.

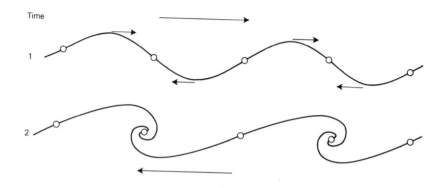

Fig. 7

But when there is a stable density gradient new vorticity
is created at the up-slope nodes, and it is destroyed at the
downslope nodes. If the density gradient is strong enough to
generate this new vorticity faster than it is transported as
just described, by the velocity field of the vorticity itself,
then the waves will not grow. This consideration gives the
usual criterion for instability for a given wavelength in
terms of the Richardson number.

It might be thought that with a stable enough density
gradient all vortex layers could be made stable. But in the
great outdoors of the atmosphere and ocean the outcome is
different.

The Richardson number is crucial. $g\beta$ (often denoted by N^2)
represents the static stability. β represents either a density
gradient in the layer or a density difference across it, and
therefore, returning to the vorticity equation again, we see
that when the layer is tilted it represents the rate of
generation of shear. The shear generated is proportional to
the time for which the layer is kept tilted, and so additional
shear is added which is proportional to β. If the shear is
$\partial U/\partial z$, then after a time τ, the Richardson number takes the
form

$$R_1 = \frac{g\beta}{(\partial U/\partial z)^2} \propto \frac{g\beta}{(\beta\tau)^2} \propto \frac{1}{\beta}$$

the constant of proportionality depending on the slope of the
layer away from horizontal. The significant factor is the
final one which shows that, ceteris paribus, the larger the
static stability the smaller will the Richardson number become
when the layer is tilted. The consequence is that under
suitable larger scale circumstances which tilt the layer, the
cause of overturning may be considered to be the original large
static stability.

When a stably stratified layer is overturned by billows the
gravitational instability inside the billows may cause a
thorough mixing so that the end result is a mixed layer between
two 'discontinuities' of density, or thin stable layers, at
which the whole sequence of events may be repeated if the layer
is again tilted.

8. MODELS OF TURBULENCE

What has been described here is not a model of turbulence
designed to facilitate a mathematical treatment or a dimen--
sional analysis, it is a statement of how the mechanism
operates and is based on the basic equations. We have not
discussed the role of viscosity, but, according to what has
already been said, the consequences of having viscosity appear
mainly at boundaries such as the surfaces of grid elements or
obstacles from which vortex sheets take off into the bulk of
the fluid. Viscosity obviously has a role in dissipating
kinetic energy, and a useful way of thinking of this is to note
that in the stretching process the eddy Reynolds number $r^2 \omega / \nu$,
remains unaltered. The time taken for the eddy to come to a
standstill because of viscosity is greatly reduced by
stretching, and the reduction in eddy size by stretching is as
important as the transfer of energy from larger sizes. Artificial models such as mixing length theories give no insight
into these matters because they prejudge every issue by their
nature.

9. PRODUCTION OF NOISE

Finally it is interesting to see why turbulence generates
noise. Again there is no pretence that we can calculate what
happens, but we can indicate how to make a lot of noise, and
perhaps how to reduce it.

When a vortex tube is stretched it is spun up so that the
pressure is lowered along its axis. This must be accompanied
by an expansion of the air inside it. The mechanism lies in
the increased centrifugal forces which expand the tube. If

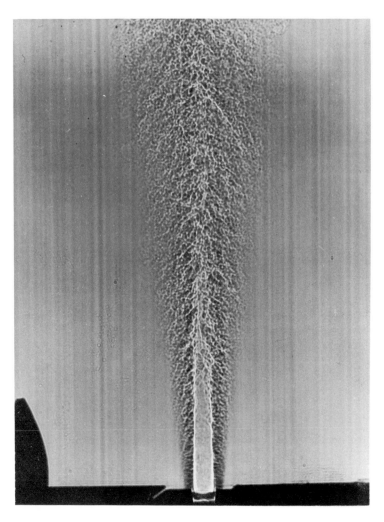

this happens quickly enough it will emit sound of audible
wavelength.

In this process we ask what happens at the ends of the tube.
In order that the stretching shall take place the ends must be
knotted into a bundle of vortex spaghetti, or the stretching
must be so rapid that the tube is disrupted. There is a
maximum speed at which pulses can be transmitted along the
tube, like a piece of elastic subjected to longitudinal waves,
and if it is stretched so rapidly that the two ends no longer
remain in communication the tube is ruptured and after the
expansion resulting from the spin-up air may enter at the open
ends and a local contraction takes place when the low pressure
centre of the tube fills up.

In shear flow the lines of fluid which are stretched most
rapidly are those inclined at $45°$ to the flow. For those who
are still sceptical or incredulous I include a shadowgraph
made by Professor Mollo-Christensen of a high speed jet in
which we can see that the spaghetti inclined at $45°$ is the most
intense and has the lowest central pressure and density.

The frequency of the noise emitted depends on the rate of
production of these "hollow" tubes, i.e. the time they take to
appear and disappear. If they burst into small bits each bit
may be said to resemble a quadrupole, but those which do not
burst would presumably emit the most intense radiation as line
monopoles in the direction of the common perpendicular to
their axes, that is at $45°$, which is where the most intense
sound is observed.

REFERENCES

Scorer, R.S. (1978) Environmental Aerodynamics, particularly
    chapters 6, 7, 8, 9. Ellis, Horwood & Halstead Press.

Scorer, R.S. (1958) Natural Aerodynamics, Ch. 6, Pergamon.

CALCULATION OF STABLY STRATIFIED SHEAR-LAYER FLOWS WITH A BUOYANCY - EXTENDED K-ε TURBULENCE MODEL

W. Rodi

(Institut für Hydromechanik, University of Karlsruhe, F.R. Germany)

ABSTRACT

A buoyancy-extended version of the k-ε turbulence model is presented which is derived by simplifying modelled transport equations for the Reynolds stresses, turbulent heat fluxes and temperature fluctuations. The buoyancy source terms in these equations are retained in the process of simplifying them to algebraic expressions. The turbulent kinetic energy k and rate of its dissipation ε appearing in the algebraic expressions are determined from transport equations for these quantities, which also contain buoyancy terms. The model is given in a form applicable to horizontal stratified shear layers, in which case the algebraic stress and flux relations can be expressed in the eddy viscosity/diffusivity form used in the k-ε model; the difference to the basic k-ε model is that two constants are now functions of buoyancy effects. The performance of the model is demonstrated by application to the following stably stratified shear layers; 2D heated surface jet discharged into stagnant ambient, stratified wall jet, vertical mixing in stratified channel flow, and plane wake in stably stratified environment. In each case, the vertical mixing and the spreading of the shear layers are reduced by the stable stratification. By reference to experimental results, the model is shown to simulate well this observed influence of stratification.

1. INTRODUCTION

Stably stratified flows occur often in the natural environment, where the stratification usually has a strong influence on the flow as well as on the diffusion of pollutants. Because of the great practical significance of stably stratified flows in the environment, engineers, meteorologists and oceanographers need practical methods for computing them, including the turbulent transport processes associated with such flows.

Those turbulence models which represent approximate methods for second-order moments range from models based on Prandtl's mixing length hypothesis to those that involve differential transport equations for the individual turbulent stresses and fluxes. A review of these models has been given by Rodi (1980). The simple mixing-length hypothesis lacks universality because it assumes that the turbulence is determined by the local mean flow and neglects transport and history effects. Hence it implies that turbulence is in local equilibrium. As it is difficult to prescribe the mixing-length distribution in situations other than simple shear-layer flows, the model is not suitable for flows with separated regions. Further, entirely empirical modifications to the model are necessary to make it applicable to buoyant flows. Unfortunately, these modifications have not proven to be of general applicability.

Energy-equation models that solve the equation for the kinetic energy of the turbulent motion, k, are superior to mixing-length models because they account in some ways for transport and history effects. Also, due to the presence of buoyancy production/destruction terms in the kinetic energy equation, the influence of stratification enters naturally into the model. However, as in mixing-length models, the length scale of turbulence has to be prescribed and suffers from the same lack of universality.

At the other end of the spectrum of models are the so-called second-order closure schemes which employ transport equations for the individual stresses and fluxes. Since they are derived from the Navier-Stokes equations and the corresponding time-dependent temperature or concentration equation, they contain terms explicitly accounting for the influence of buoyancy forces on the turbulent stresses or fluxes. Even though the exact forms of these equations cannot be used directly but require the introduction of model approximations, they form the most realistic basis for taking into account the effects of buoyancy on the turbulent stresses and fluxes. The resulting model is however rather complex which makes it less suitable for solving practical problems.

At an intermediate level two-equation models, which provide only differential equations for the velocity and length scale of the turbulent motion, have been found to be a fairly success-ful compromise between universality and simplicity. Especially the k-ε model has been shown to work quite well for many different flows not controlled by buoyancy forces (Rodi, 1980, 1984). The basic version of the k-ε model allows for some influence of buoyancy forces through buoyancy production or destruction terms in the kinetic energy equation and in the length-scale-determining equation for the dissipation rate ε. However, this model does not account for the influence of stratification on the ratios of the stresses to each other, the ratios of the stresses to the turbulent kinetic energy (structure parameters) and on the ratio of turbulent heat or mass flux to turbulent momentum flux.

To account for these effects of stratification and for the effects of nearby boundaries, an extended version of the k-ε model is described in the present paper that is derived by simplifying the more complex, but still approximate transport-equation model of Gibson and Launder (1978). The simplification results in algebraic relations for the individual turbulent stresses and fluxes which involve the quantities k and ε. For the stratified shear layers to be considered in this paper, the algebraic relations can be written in the eddy viscosity/diffusivity form used in the k-ε model, the difference to the basic model is that two of the constants are replaced by functions of the stratification.

A detailed derivation of the buoyancy-extended k-ε model is given in Hossain and Rodi (1982), where applications of the model to various vertical buoyant jets are presented. In the present paper, the buoyancy-extended k-ε model is applied to several horizontal shear layers with stable stratification, namely to two-dimensional heated surface jets discharged into stagnant ambients, stably stratified wall jets, stably stratified open channel flow and a plane wake in stratified environment. The calculated results are compared with experiments in order to show that the model simulates realistically the reduction of mixing and spreading caused by a stable stratification.

2. MATHEMATICAL MODEL

2.1 Mean-Flow Equations

The distribution of velocity and temperature or concentration in horizontal shear layers is governed by the following equations

$$\frac{\partial U}{\partial x} + \frac{\partial V}{\partial y} = 0 \tag{1}$$

$$U \frac{\partial U}{\partial x} + V \frac{\partial U}{\partial y} = - \frac{\overline{\partial uv}}{\partial y} - \frac{1}{\rho} \frac{\partial P}{\partial x} \qquad (2)$$

$$U \frac{\partial T}{\partial x} + V \frac{\partial T}{\partial y} = - \frac{\overline{\partial vT'}}{\partial y} \qquad (3)$$

where equation (1) is the continuity equation, (2) the x-momentum equation and (3) the temperature transport equation (which also represents the transport equation for concentration when T stands for concentration); the symbols are defined in Fig. 1. As is usual for thin shear layers, the normal-stress term was neglected in the momentum equation (2) and the longitudinal heat-flux term in the temperature equation (3). Further, molecular transport terms have been neglected so that the equations are not valid in the viscous sub-layer near walls. When the equations are applied to flows near walls in the present paper, this layer is bridged by so-called wall functions as explained in the section on boundary conditions below. In stratified situations, the longitudinal pressure gradient appearing in the momentum equation (2) is not strictly zero, as can be seen by integrating the hydrostatic pressure condition approximating the vertical momentum equation. However, in most cases the pressure gradient caused by buoyancy forces is negligible compared with the other terms; only when the buoyancy forces are of comparable magnitude with the inertial forces (at relatively large Richardson numbers) is the pressure-gradient term of significance. This term was therefore neglected in most of the application examples given below; it was retained only in the case of the heated surface jet, where the actual form used for $\partial P/\partial x$ is given in the section where this example is discussed.

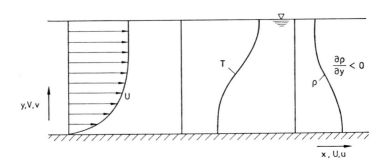

Fig. 1 Flow configuration

The momentum equation contains the turbulent shear stress $\overline{uv}$ and the temperature equation the vertical turbulent heat flux $\overline{vT'}$ as unknowns. These turbulent correlations must be determined with the aid of a turbulence model in order to close the system of equations, and the next sub-sections introduce such models.

2.2 Basic k-ε Model

The k-ε model relates the shear stress $\overline{uv}$ to the local rate of strain and the turbulent heat flux $\overline{vT'}$ to the local temperature gradient via the following eddy viscosity and diffusivity relations

$$- \overline{uv} = v_t \frac{\partial U}{\partial y} \quad , \quad - \overline{vT'} = \Gamma_t \frac{\partial T}{\partial y} \tag{4}$$

The model assumes that the local state of turbulence can be characterized by two parameters, namely the turbulent kinetic energy k and the rate of its dissipation, ε, and relates the eddy viscosity v_t and the eddy diffusivity Γ_t to these parameters via the following relations

$$v_t = c_\mu \frac{k^2}{\varepsilon} \quad , \quad \Gamma_t = \frac{v_t}{\sigma_t} \tag{5}$$

where c_μ and σ_t are empirical constants, the latter one being the turbulent Prandtl/Schmidt number. The distributions of k and ε in the shear layer are obtained from modelled transport equations for these quantities, which read for horizontal buoyant shear layers

$$U \frac{\partial k}{\partial x} + V \frac{\partial k}{\partial y} = \frac{\partial}{\partial y} \left[\frac{v_t}{\sigma_k} \frac{\partial k}{\partial y} \right] + \underbrace{v_t \left(\frac{\partial U}{\partial y} \right)^2}_{P} + \underbrace{\beta g \frac{v_t}{\sigma_t} \frac{\partial T}{\partial y}}_{G} - \varepsilon \tag{6}$$

$$U \frac{\partial \varepsilon}{\partial x} + V \frac{\partial \varepsilon}{\partial y} = \frac{\partial}{\partial y} \left[\frac{v_t}{\sigma_\varepsilon} \frac{\partial \varepsilon}{\partial y} \right] + c_{1\varepsilon} \frac{\varepsilon}{k} [P + (1-c_{3\varepsilon})G] - c_{2\varepsilon} \frac{\varepsilon^2}{k} \tag{7}$$

where $c_{1\varepsilon}$, $c_{2\varepsilon}$, $c_{3\varepsilon}$, σ_k and σ_ε are further empirical constants.
P is the production of kinetic energy by the interaction of
shear stress and mean velocity gradient and read originally
$-\overline{uv}\partial U/\partial y$; G is the production or destruction of kinetic energy
k by buoyancy forces and read originally $\beta g\overline{vT'}$. With the eddy
viscosity/diffusivity relations (4) introduced, these original
expressions result into the terms given in equation (6). The
ε-equation given here for horizontal buoyant shear layers is a
special form of the general ε-equation suggested by Rodi
(1979, 1980) which has as source term the expression

$$c_{1\varepsilon}\,\frac{\varepsilon}{k}\,(P + G)(1 + c_{3\varepsilon}R_f)$$

in which R_f is the flux Richardson number based on the buoyancy
production of the lateral fluctuations $\overline{v^2}$, i.e.
$R_f = -1/2\mathrm{Prod}_{\overline{v^2}} / (P+G)$. For horizontal buoyant layers, the
production term in the $\overline{v^2}$-equation will be given below in
equation (12). In this case, all buoyancy production goes into
the component $\overline{v^2}$ so that $\mathrm{Prod}_{\overline{v^2}} = 2G$ and $R_f = -G/(G+P)$, which

then yields the special form given in the ε-equation (7). The
standard constants used in the k-ε model are listed in Table 1;
they are taken from Rodi (1980).

Table 1

Constants in k-ε model

c_μ	$c_{1\varepsilon}$	$c_{2\varepsilon}$	$c_{3\varepsilon}$	σ_k	σ_ε
0.09	1.44	1.92	0.8	1.0	1.3

Apart from the value of $c_{3\varepsilon}$, these constants are well established and seem to work well in many different flow situations.
For the constant $c_{3\varepsilon}$ in the buoyancy term of the ε-equation,
quite different values were reported in the literature. The
value of 0.8 was determined in the course of the heated-surface-
jet study reported below. According to equation (7), this value
indicates that only a small part of the buoyancy production is
active to produce ε (or to destroy ε in the case of stable

stratification). Reasonable calculations can therefore also
be obtained for the heated surface jet when the buoyancy term
is neglected altogether in the ε-equation ($c_{3\varepsilon}$ = 1.0), as was
done by Gibson and Launder (1976). This finding was confirmed
by Goussebaile and Viollet (1982) who calculated successfully
a stably stratified mixing layer with $c_{3\varepsilon}$ = 1. On the other
hand, Svensson (1980) found that the development of stably
stratified mixed layers driven by a surface shear stress can be
simulated best with $c_{3\varepsilon}$ = 0.4. In all the free shear layer
situations considered so far, the buoyancy term G was a sink
term. When G is a source term as in an unstably stratified
mixing layer, apparently $c_{3\varepsilon}$ = 0 is required (implying that
the same multiplying constant for shear and buoyancy production
terms in the ε-equation is used). This was confirmed by
Hossain and Rodi (1982) for the case of vertical buoyant jets,
where G is also positive. In stratified wall boundary layers
with heat flux to or from the wall again quite different values
for $c_{3\varepsilon}$ are needed. Betts and Haroutunian (1983) report values
of 2.15 and -0.8 for stably and unstably stratified atmospheric
boundary layers, respectively.

    From the above discussion it is clear that the ε-equation
(7) with a constant value of 0.8 for $c_{3\varepsilon}$ is suitable only for
certain stratified shear layers, in particular it does not
reproduce the observed behaviour of stratified wall boundary
layers. However, this ε-equation works quite well for all the
stably stratified free shear layers considered below.

    In non-buoyant flows under local-equilibrium conditions, the
empirical constant c_μ can be shown to be equal to the square of
the structure parameter $\overline{uv}/k$. This parameter was found to be
reduced significantly by stable stratification, a fact that is
not accounted for in the basic k-ε model which assumes c_μ to be
constant. Further, the turbulent Prandtl/Schmidt number σ_t has
also been measured to depend on buoyancy effects and is not
really a constant under stratified conditions. In order to
allow for the dependence of c_μ and σ_t on stratification
effects, the k-ε model is extended in the following subsection
where buoyancy-dependent functions for these two parameters
are derived.

2.3 Buoyancy-Extended k-ε Model

The derivation of the buoyancy-extended k-ε model starts
from the second-order-closure scheme of Gibson and Launder
(1978) which uses modelled transport equations for the
individual Reynolds stresses $\overline{u_i u_j}$, turbulent heat fluxes $\overline{u_i T'}$
and temperature fluctuations $\overline{T'^2}$. The general forms of these
equations are given by the following equations

$$U_\ell \frac{\partial \overline{u_i u_j}}{\partial x_\ell} = \underbrace{\text{Diff}}_{\overline{u_i u_j}} - \underbrace{\overline{u_i u_\ell} \frac{\partial U_j}{\partial x_\ell} - \overline{u_j u_\ell} \frac{\partial U_i}{\partial x_\ell}}_{} - \underbrace{\beta(g_i \overline{u_j T'} + g_j \overline{u_i T'})}_{}$$

| convective transport | diffusive transport | P_{ij}=stress production | G_{ij}=buoyancy production |

$$- c_1 \frac{\varepsilon}{k}(\overline{u_i u_j} - \frac{2}{3}\delta_{ij}k) + c_1'(\overline{u_n^2}\delta_{ij} - \frac{3}{2}\overline{u_n u_i}\delta_{nj} - \frac{3}{2}\overline{u_n u_j}\delta_{ni})f\left(\frac{L}{x_n}\right)$$

$$\underbrace{- c_2(P_{ij} - \frac{2}{3}\delta_{ij}P)}_{} + c_2'(\pi_{nn}\delta_{ij} - \frac{3}{2}\pi_{ni}\delta_{nj} - \frac{3}{2}\pi_{nj}\delta_{ni})f\left(\frac{L}{x_n}\right)$$

$$\pi_{ij}$$

$$- c_3(G_{ij} - \frac{2}{3}\delta_{ij}G)$$

$$\underbrace{}_{}$$

pressure-strain   (8)

$$\underbrace{- \frac{2}{3}\varepsilon\delta_{ij}}_{}$$

dissipation

$$U_\ell \frac{\overline{\partial u_i T'}}{\partial x_\ell} = \text{Diff}_{\overline{u_i T'}} - \overline{u_i u_j} \frac{\partial T}{\partial x_i} - \overline{u_j T'} \frac{\partial U_i}{\partial x_j} - \beta g \overline{T'^2}$$

| convection transport | diffusive transport | mean-field production | buoyancy production |

$$- c_{1T} \frac{\varepsilon}{k} \overline{u_i T'} - c_{1T'} \frac{\varepsilon}{k} \overline{u_n T'} \delta_{in} f\left(\frac{L}{x_n}\right)$$

$$- c_{2T} \overline{u_\ell T'} \frac{\partial U_i}{\partial x_\ell} - c_{3T} \beta g \overline{T'^2} \tag{9}$$

pressure-scrambling

$$U_\ell \frac{\overline{\partial T'^2}}{\partial x_\ell} = \text{Diff}_{\overline{T'^2}} - 2 \overline{u_j T'} \frac{\partial T}{\partial x_j} - \frac{1}{R} \frac{\overline{T'^2}}{k} \varepsilon \tag{10}$$

| convective transport | diffusive transport | mean-field production | molecular destruction |

The physical signficance of the individual terms is indicated in the equations. Exact forms of these equations can be derived which contain automatically buoyancy terms. However, in order to obtain a closed scheme, model assumptions had to be introduced into the exact equations about the diffusion, viscous destruction and pressure strain/scrambling terms. The latter represent correlations between the fluctuating pressure and the fluctuating velocity and temperature gradients and are of particular importance in the transport equations. Hence, the most important aspect of the Gibson-Launder model is the suggestion for the pressure-strain/scrambling term which involves a buoyancy contribution and also a surface correction accounting for the damping of lateral fluctuations due to the presence of a solid or free surface. The terms involving f in the pressure strain/scrambling model are the surface-proximity corrections, where f is an empirical damping function which is to reduce the effect of the surface correction with distance from the surface. In these correction terms, the subscript n

$$U\frac{\partial \overline{uv}}{\partial x} + V\frac{\partial \overline{uv}}{\partial y} = \text{Diff}_{\overline{uv}} - \overline{v^2}\frac{\partial U}{\partial y} + \beta g\overline{uT'} - c_1\frac{\epsilon}{k}\overline{uv}\left(1 + \frac{3}{2}\frac{c_1'}{c_1}f\right) + c_2\left(1 - \frac{3}{2}c_2'f\right)\overline{v^2}\frac{\partial U}{\partial y} - c_3\beta g\overline{uT'} \tag{11}$$

$$U\frac{\partial \overline{v^2}}{\partial x} + V\frac{\partial \overline{v^2}}{\partial y} = \text{Diff}_{\overline{v^2}} + 2\beta g\overline{vT'} - \frac{2}{3}\epsilon - c_1\frac{\epsilon}{k}\left[\left(1 + 2\frac{c_1'}{c_1}f\right)\overline{v^2} - \frac{2}{3}k\right] \tag{12}$$

$$U\frac{\partial \overline{uT'}}{\partial x} + V\frac{\partial \overline{uT'}}{\partial y} = \text{Diff}_{\overline{uT'}} - \overline{uv}\frac{\partial T}{\partial y} - \overline{vT'}\frac{\partial U}{\partial y} - \frac{2}{3}c_2\left(1 - 2c_2'f\right)\overline{uv}\frac{\partial U}{\partial y} - \frac{4}{3}c_3\beta g\overline{vT'} - c_{1T}\frac{\epsilon}{k}\overline{uT'} + c_{2T}\overline{vT'} \tag{13}$$

$$U\frac{\partial \overline{vT'}}{\partial x} + V\frac{\partial \overline{vT'}}{\partial y} = \text{Diff}_{\overline{vT'}} - \overline{v^2}\frac{\partial T}{\partial y} + \beta g\overline{T'^2} - \left(c_{1T} + c_{1T}'f\right)\frac{\epsilon}{k}\overline{vT'} - c_{3T}\beta g\overline{T'^2} \tag{14}$$

$$U\frac{\partial \overline{T'^2}}{\partial x} + V\frac{\partial \overline{T'^2}}{\partial y} = \text{Diff}_{\overline{T'^2}} - 2\overline{vT'}\frac{\partial T}{\partial y} - \frac{\epsilon}{kR}\overline{T'^2} \tag{15}$$

| convection | diffusion | mean-field production | buoyancy prod. | viscous destr. | pressure-strain/scrambling |

denotes the direction normal to the surface so that $\overline{u_n^2}$ are the lateral fluctuations and x_n is the distance from the surface.

Gibson and Launder (1978) assume f to be a linear function of L/x_n, where L is the local length scale of the turbulence. The actual function used will be presented below. Further, the models for the diffusion terms in equations (8) to (10) are not given explicitly, as further simplifying assumptions are introduced for the diffusion shortly.

Special forms of the transport equations valid for two-dimensional horizontal buoyant shear layers are obtained by neglecting all velocity and temperature gradients except $\partial U/\partial y$ and $\partial T/\partial y$. For these flows, the prime interest lies in the equations for the turbulent shear stress $\overline{uv}$ and for the vertical turbulent heat flux $\overline{vT'}$, but since in these equations there appear also the vertical velocity fluctuations $\overline{v^2}$, the longitudinal heat flux $\overline{uT'}$ and the temperature fluctuations $\overline{T'^2}$, equations are given also for these correlations. The total set is compiled in equations (11) to (15).

Together with the k- and ε-equations and an explicit model for the diffusion terms, these equations form a second-order closure scheme. However, the equations are used here only as a starting point for the derivation of simpler algebraic equations for the turbulent stresses and heat fluxes. As it is the convection and diffusion terms that make the transport equations differential equations, simplifying model approximations are introduced for these terms. It is important to note, however, that all source terms containing the buoyancy and surface damping effects are retained during the process of simplification. As $\overline{v^2}$ is a component of the total turbulent kinetic energy k, its combined convective and diffusive transport is assumed proportional to the transport of k, which according to equation (6) is:

$$(\text{Conv.-Diff.})_{\overline{v^2}} = \frac{\overline{v^2}}{k} (\text{Conv.-Diff.})_k = \frac{\overline{v^2}}{k} (P + G - \varepsilon). \quad (16)$$

For the other turbulence correlations, convection and diffusion are simply neglected on the assumption that turbulence is nearly in local equilibrium. Such a neglect would in general also be a reasonable approximation for $\overline{v^2}$, but this is not true near free surfaces where the velocity gradient and hence the production of turbulence are zero so that the diffusive transport is necessary to balance the dissipation at the surface. Refined treatment of $\overline{uv}$ and $\overline{vT'}$ near the free surface is not important because these quantities go to zero at such a surface. On the other hand, $\overline{uT'}$ and $\overline{T'^2}$ have only an indirect influence on the correlations of prime interest and a rather crude model for these terms is therefore sufficient.

With these assumptions on the convection and diffusion terms for the various dependent variables of equations (11) to (15), all these equations can be reduced to algebraic expressions for the individual correlations. The resulting expressions for $\overline{uv}$ and $\overline{vT'}$ can be written in the eddy viscosity/diffusivity form of equation (4) with the eddy viscosity and diffusivity expressed through equation (5). This means that the parameters c_μ and σ_t now depend on stratification and on surface damping:

$$c_\mu = \omega \frac{\overline{v^2}}{k} , \quad \sigma_t = \frac{\omega}{\alpha} , \tag{17}$$

$$\text{with} \quad \omega = \frac{1 - c_2 + \frac{3}{2} c_2 c_2 'f}{c_1 + \frac{3}{2} c_1 'f} \cdot \frac{1 - \dfrac{1 - c_3}{1 - c_2 + \frac{3}{2} c_2 c_2 'f} \cdot \dfrac{1 - c_{2T}}{c_{1T}} \alpha B}{1 + \dfrac{1 - c_3}{c_1 + \frac{3}{2} c_1 'f} \cdot \dfrac{1}{c_{1T}} B} , \tag{18}$$

$$\alpha = \frac{1}{c_{1T} + c_{1T}' f + 2(1 - c_{3T}) RB} , \quad B = \beta g \frac{k^2}{\varepsilon^2} \frac{\partial T}{\partial y} , \tag{19}$$

$$\frac{\overline{v^2}}{k} = \frac{2}{3} \cdot \frac{c_1 - 1 + \dfrac{P+G}{\varepsilon}(c_2 - 2c_2 c_2 'f) + \dfrac{G}{\varepsilon}(3 - c_2 - 2c_3 + 2c_2 c_2 'f)}{c_1 + 2c_1 'f + \dfrac{P + G}{\varepsilon} - 1} \tag{20}$$

The correlations $\overline{uT'}$ and $\overline{T'^2}$ originally appearing in this set of algebraic expressions for horizontal shear layers have been eliminated with the aid of the simplified forms of equations (13) and (15). They are not further needed in this model. The diffusion constants σ_k and σ_ε in the k- and ε-equations are also replaced by buoyancy-dependent functions for reasons explained in detail in Hossian and Rodi (1980):

$$\sigma_k = \frac{\omega}{c_k} \qquad \sigma_\varepsilon = \frac{\omega}{c_\varepsilon} \; . \tag{21}$$

For the constants appearing in the above buoyancy extensions of the k-ε model, the following values have been adopted from Gibson and Launder (1978):

Table 2

Constants in buoyancy-extended k-ε model

c_k	c_ε	c_1	c_2	c_3	c_1'	c_2'	c_{1T}	c_{2T}	c_{3T}	$c_{1T'}$	R
0.24	0.15	2.2	0.55	0.55	0.5	0.3	3.0	0.5	0.5	0.5	0.8

Gibson and Launder (1978) considered only flows near walls and introduced the surface damping function f for such cases to be a linear function of $L/y = k^{3/2}/(\varepsilon y)$. In this paper, the model is also applied to free surface flows so that free surface damping must be considered too. Here it is assumed that the free-surface-damping function f_s can be superposed to the wall-damping function f_w resulting in

$$f = \underbrace{\frac{k^{3/2}}{c_w \varepsilon} \cdot \frac{1}{y}}_{f_w} + \underbrace{\frac{k^{3/2}}{c_w \varepsilon} \frac{1}{\bar{y} + 0.04 \, k^{3/2}/\varepsilon_s}}_{f_s} \tag{22}$$

where y is the distance from the wall and $\bar{y}$ the distance from the free surface. Both components assume basically a linear relationship between f and the ratio of length scale to distance from the surface, but for the free surface damping function f_s it is assumed that the length scale goes to zero at a virtual origin somewhat above the surface. The function (22) was developed by Hossain (1980) by reference to open-channel-flow data. In a more recent study, Gibson (1983) has found that simply using the distance from the free surface $\bar{y}$ (i.e. neglecting the shift to a virtual origin above the surface) does not cause any noticeable loss in accuracy of the

results. The constant c_w in (22) was adjusted in such a way
that f is equal to unity in the local-equilibrium near-wall
region. This choice yields $c_w = \kappa/c_{\mu 1}^{3/4}$, where κ is the
von Kármán constant in the logarithmic law of the wall and $c_{\mu 1}$
is the value of c_μ according to equation (17) at the first
numerical grid point away from the wall. Near free surfaces,
the function f goes to larger values of the order of 7, as is
shown in Hossain (1980) and Celik et al. (1982).

Gibson and Launder (1978) tested extensively a version of
the above algebraic stress/flux model in which local equilibrium was also assumed for $\overline{v^2}$, that is in equation (20)
(P+G)/ε was put equal to unity. The ratios of the individual
stresses and fluxes to each other can then be written as
functions of the flux Richardson number -G/P and the surface
damping function f. Gibson and Launder show that for unstratified local-equilibrium shear layers the model yields the
correct stress and flux ratios for both free shear layers
(f = 0) and near wall layers (f = 1). In particular, the model
simulates well the difference between the two types of layers
and hence the wall effect via the wall correction terms
involving the function f. This is also true for the turbulent
Prandtl/Schmidt number which changes from a value of 0.67 for
free layers to 0.92 for near-wall layers, in accordance with
experiments. Gibson and Launder also applied their local
equilibrium model to stratified flow situations. Remote from
walls, the ratio of vertical to longitudinal fluctuations,
$\overline{v^2}/\overline{u^2}$, decreases with increasing stable stratification, in
quantitative agreement with experiments. The opposite trend
has been observed in stably stratified boundary layers near
walls and this trend is reproduced by the model when the
influence of stratification on the turbulent length scale is
accounted for by use of the Monin-Oboukhov relation. This
reduces the length scale $L \propto k^{3/2}/\varepsilon$ under stable stratification
and hence the wall damping function f according to relation
(22). As a consequence, there is less damping of the vertical
fluctuations $\overline{v^2}$ due to the presence of the wall and hence the
fluctuations increase under stable stratification. It should
be mentioned here that the ε-equation (7) does not truthfully
reproduce the influence of stable stratification on the length
scale suggested by the Monin-Oboukhov relation and hence is
not entirely satisfactory for stratified situations with heat
flux near the wall. In the present paper, applications are
presented only to situations with an adiabatic wall. The task

of modifying the ε-equation in such a way that it would reproduce the Monin-Oboukhov relation is certainly worth pursuing.

Gibson and Launder (1978) have shown further that, in
agreement with data, the turbulent Prandtl/Schmidt number σ_t
decreases somewhat in boundary layers under the influence of
stable stratification but, what is more important for the situations considered here, increases considerably (by a factor of
the order of 2) in free shear layers.

3. APPLICATION OF THE MODEL

In this section, the model discussed above is applied to
two-dimensional heated surface jets discharged into stagnant
water, stably stratified wall jets, stably stratified open
channel flow and a plane wake in stratified environment. The
individual flow situations are introduced in greater detail
below. The model equations, which are all two-dimensional
parabolic boundary-layer equations, have been solved numerically with the marching-forward finite-difference method of
Patankar and Spalding (1970). Initial and boundary conditions
need to be specified in each case; the latter are discussed in
the next subsection while the initial conditions are problem-
dependent and will be given in the subsections on the individual flow situations.

3.1 Boundary Conditions

In the flow situations considered here, three types of
boundaries occur, namely walls, free surfaces and boundaries
between shear layer and ambient. Specification of conditions
is simplest for the last type of boundary: the ambient velocity
and temperature are prescribed and the ambient is supposed to
be free from turbulence so that k and ε are set to zero.

At solid walls, the equations are not integrated through
the viscous sublayer near the wall; instead, this layer is
bridged by adopting the wall-function approach described in
Launder and Spalding (1974) which relates the velocity at the
first grid point just outside the viscous sublayer to the wall
shear stress τ_w via the logarithmic law of the wall

$$\frac{U}{U_\star} = \frac{1}{\kappa} \ln \left(\frac{yU_\star}{\nu} E \right)$$

(23)

where $U_* = (\tau_w/\rho)^{1/2}$ is the friction velocity, κ is the von Kármán constant and E a roughness coefficient. It is further assumed that, at this point, local equilibrium prevails which leads to the following conditions for k and ε to be applied at this near wall point:

$$k = \frac{U_*^2}{\sqrt{c_\mu}}, \quad \varepsilon = \frac{U_*^3}{\kappa y}. \qquad (24)$$

Walls are considered adiabatic in the application examples so that the boundary condition for the temperature is $\partial T/\partial y = 0$

At the free surface, symmetry conditions (i.e. zero gradients) are adopted for all dependent variables but the dissipation ε. The surface value ε_s is assumed to be related to the surface value of the turbulent kinetic energy, k_s, and to a typical length scale of the shear layer, for which the layer depth h is taken:

$$\varepsilon_s = 5.4 \frac{k_s^{3/2}}{h} \qquad (25)$$

The empirical constant in this relation was determined by Hossain (1980) from channel flow data for the kinetic energy. The boundary condition (25) is based on the assumption that the macro-length scale of turbulence, $L \propto k^{3/2}/\varepsilon$ is reduced by the presence of the free surface. In fact, relation (25) does reduce the value of the length scale below the one that would result from the symmetry condition $\partial \varepsilon/\partial y = 0$. As a consequence, the eddy viscosity is reduced near the surface leading in open channel flow to the observed parabolic vertical distribution of the eddy viscosity.

3.2 2D Heated Surface Jet

The first application example concerns the flow situation sketched in Fig. 2a which results from discharging a jet of heated water horizontally at the surface into a tank with stagnant colder water. For the calculations, the receiving water body is assumed to be infinitely large so that there is no downstream control on the jet flow and this is of the boundary-layer type. In cases with downstream control, e.g. due to the limited size of the receiving tank, the jet ends in an internal hydraulic jump which is followed by a two-layer

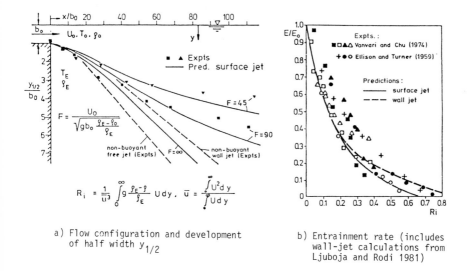

a) Flow configuration and development of half width $y_{1/2}$

b) Entrainment rate (includes wall-jet calculations from Ljuboja and Rodi 1981)

Fig. 2   2D heated surface jet, from Hossain (1980)

flow. As internal jumps involve roller regions, the flow is no longer of the boundary-layer type and cannot be calculated with the present model. Calculations of this flow phenomenon were obtained by Leschziner (1979) who used the more general elliptic equations and solved them with an iterative finite difference method.

In the case of heated surface jets, buoyancy has two effects on the jet spreading. One is to generate excess hydrostatic pressure forces (baroclinic forces) which cause the pressure gradient $\partial P/\partial x$ in the momentum equation to assume finite values. For the 2D buoyant surface jet, this pressure gradient can be related to the density profile by

$$- \frac{\partial P}{\partial x} = \frac{\partial}{\partial x} \left[\int_y^\infty (\rho - \rho_E) \, g \, dy \right] \qquad (26)$$

where y is zero at the surface and increases in the downward direction. This relation results from calculating the vertical pressure distribution with the aid of the hydrostatic pressure approximation. The resulting pressure relation involves the superelevation η caused by the buoyancy forces so that the surface slope $\partial\eta/\partial x$ appears originally in the expression for the pressure gradient $\partial P/\partial x$. The surface slope is eliminated by the condition that at large depths ($y \to \infty$) no horizontal motion is induced (i.e $\partial P/\partial x = 0$ at $y \to \infty$); this yields the excess hydrostatic pressure term (26) used in the calculations presented here. Excess hydrostatic pressure effects are not of major significance for 2D surface jets; they are much more important in the case of a 3D heated surface jet where they cause a very strong lateral surface spreading of the warm water like that of an oil film (see e.g. McGuirk and Rodi, 1979). The second effect of buoyancy is to influence turbulence, and this is of much greater importance in the case of a 2D jet. The discharged heated water sets up a stable stratification which damps the turbulence and consequently reduces the jet entrainment and spreading until turbulence finally collapses and the warmer water floats on top of the colder water in a pronounced two-layer situation. Two experimental investigations were carried out on the 2D surface jet: Ellison and Turner (1959) measured only the entrainment while Vanvari and Chu (1974) also measured velocity and temperature profiles as well as the longitudinal velocity fluctuations. The latter experimenters took special care that no internal hydraulic jump developed while Ellison and Turner carried out measurements only shortly after the onset of the jet before gravity waves were reflected at the opposite wall of the tank, leading to the formation of an internal hydraulic jump. The flow was probably not entirely steady in this initial period when the measurements were taken. Ellison and Turner plotted the measured entrainment rate E ($= \partial\int_0^\infty Udy/\partial x$) versus the Richardson number $\overline{Ri}$, which is defined in Fig. 2a as the cross-sectional buoyancy flux divided by a characteristic velocity scale $\bar{u}$. $\overline{Ri}$ characterises the influence of buoyancy at a particular jet cross section: it has an initial value at the jet exit and increases monotonically until turbulence collapses and the entrainment ceases. Vanvari and Chu tried to compare their entrainment measurements with those of Ellison and Turner and found that their entrainment goes to zero at much smaller $\overline{Ri}$-values than in the case of Ellison and Turner. However, in their evaluation of the Richardson number $\overline{Ri}$ they used as velocity scale the surface velocity while Ellison and Turner used an average velocity defined in Fig. 2a. Hossain (1980) has reevaluated Vanvari and Chu's data using Ellison and Turner's definition of the velocity scale; the reevaluated data are plotted together with Ellison and Turner's data in Fig. 2b and show fairly reasonable agreement.

The marching solution was started at the jet exit with prescribed uniform velocity and temperature profiles that yielded
the densimetric Froude numbers F (defined in Fig. 2a) investigated in the experiments of Vanvari and Chu. As initial conditions for k and ε, small uniform values were used at the exit
yielding a small eddy viscosity. Fig. 2a compares the calculated jet spreading (half-width of the jet) for three different
exit Froude numbers with the experimental results. The reduction of spreading by decreasing the exit Froude number is simulated well by the model. The figure shows further that the
calculated spreading of the non-buoyant surface jet (F = ∞) is
smaller than that of the non-buoyant free jet. The difference
results from the inclusion of the surface-damping correction in
the pressure-strain model which reduces the vertical fluctuations $\overline{v^2}$ according to relation (20), and as a consequence also
the parameter c_μ, the shear stress $\overline{uv}$ and hence the jet spreading. This is now closer to that of a wall jet, a result which
seems plausible in view of the fact that lateral fluctuations
are damped both by a wall and a free surface. Ljuboja and Rodi
(1980) have shown that the spreading of a wall jet is predicted
correctly with a model very similar to the present one. Fig.
2b shows the entrainment rate E, non-dimensionalised with its
value for non-buoyant jets E_0, as a function of the Richardson
number $\overline{Ri}$. The strong reduction of entrainment with increasing
Richardson number is simulated very well by the model. It
should be mentioned here that the experimental data shown in
Fig. 2b were used to determine the empirical constant $c_{3\varepsilon}$.
However, the entrainment results are not very different when
the buoyancy term in the ε-equation is neglected altogether
(amounting to $c_{3\varepsilon} = 1$), a practice adopted by Gibson and Launder
(1976) in their calculations of the buoyant surface jet. It
should further be mentioned that neglecting the excess hdyro-
static pressure term in the momentum equation has a noticeable
effect only for $\overline{Ri} > 0.4$ (see Hossain 1980).

   Fig. 3a displays the calculated velocity profiles at three
different $\overline{Ri}$-numbers. For the situation without buoyancy
($\overline{Ri} = 0$), the profile is the similarity profile which agrees
well with experimental results. The influence of stable
stratification is to make the velocity profile fuller, and this
trend is predicted in good agreement with the data. Fig. 3b
shows the calculated density profile (basically T-profile) for
$\overline{Ri} = 0.4$ and 0.7. The measured profile for $\overline{Ri} = 0.04$ is also
included and indicates a fairly linear behaviour which was
also observed in 3D surface jets (see McGuirk and Rodi, 1979).
This trend is not reproduced very well by the model and a
further examination of this phenomenon is necessary. Fig. 3b
also includes the calculated distributions of the turbulent

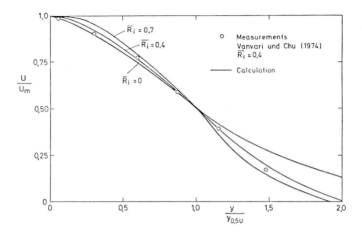

a) Velocity profiles

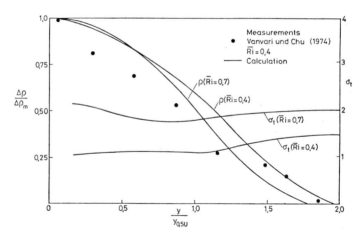

b) Profiles of ρ and σ_t

Fig. 3   Velocity and density profiles in 2D heated surface
         jets, from Hossain (1980)

Prandtl number σ_t for the two Richardson numbers and illustrates that the level of σ_t rises from 1.3 for $\overline{Ri}$ = 0.4 to 1.8 for $\overline{Ri}$ = 0.7.

Fig. 4 shows the calculated decay of the relative maximum turbulent kinetic energy at any cross section and also of the maximum temperature fluctuations as $\overline{Ri}$ increases. Beyond $\overline{Ri} \cong 0.7$, both quantities drop sharply, indicating the collapse of turbulence. Unfortunately, due to lack of data a quantitative comparison with measurements is not possible but the measured longitudinal fluctuations $\overline{u'^2}$ in Fig. 4 indicate that the trend is predicted correctly.

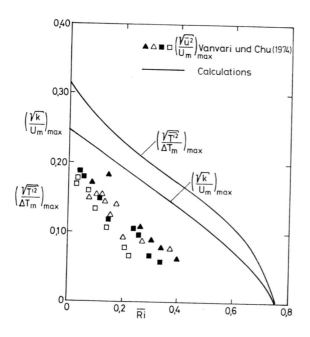

Fig. 4   Decay of maximum k and $\overline{T'^2}$ with $\overline{Ri}$ in heated surface jets, from Hossain (1980)

3.3 Stratified Wall Jets

The next application example concerns jets discharged from a slot with width b_o along a horizontal wall at a temperature

T_O that is lower than the temperature T_E of the stagnant ambient above the jet. This flow situation is very similar to the surface-jet situation discussed above, as again a stable stratification is set up which damps the turbulence and reduces the entrainment and jet spreading. Ljuboja and Rodi (1981) used a model very similar to the one introduced above to calculate buoyant wall jets for three different exit Froude numbers (definition as given in Fig. 2a) and also the limiting case of a non-buoyant wall jet ($F = \infty$). For this case, Ljuboja and Rodi (1980) had shown in an earlier paper that the model yields good agreement with experimental results. Unfortunately, no experimental data were available for buoyant wall jets in stagnant surroundings.

The calculations were started at the jet exit with uniform velocity and temperature profiles and low values for the turbulence quantities k and ε. The development of the jet was then calculated up to an axial station of $x/b_O = 300$. Fig. 5 shows the predicted development of the half-width of the velocity field $y_{1/2}$ and of the temperature field $y_{1/2T}$ and the decay of the maximum velocity U_m. It can be seen that the jet spreading and the decay of U_m are reduced significantly by the stable stratification. For the lowest value of F, the jet spreads very little at the furthest downstream station of $x/b_O = 300$.

It is further worth noting that the temperature field spreads faster than the velocity field for the non-buoyant case, which reflects the fact that the turbulent Prandtl number is smaller than unity. When the Froude number is reduced however the temperature field spreads increasingly less than the velocity field. This is due to the fact that the turbulence model predicts the turbulent Prandtl number to rise above unity for stably stratified situations, as was shown in Fig. 3b. The reduced spreading is due to a reduction of the entrainment of ambient fluid into the jet. The predicted influence of stratification on the entrainment is compared with that in the surface jet in Fig. 2b, where the entrainment rate E, non-dimensionalised with its value for the non-buoyant wall jet E_O, is plotted versus the Richardson number $\overline{Ri}$ defined in Fig. 2a. In this way of plotting, the results of the calculations for the three F-values collapse on a single curve, except for the developing region near the exit which is excluded from Fig. 2b. Within the experimental scatter, the single curve agrees well with the data obtained in the closely related situation of a heated surface jet and is also not very different from Hossain's (1980) calculations for that case. This illustrates convincingly that the two flows are indeed very similar.

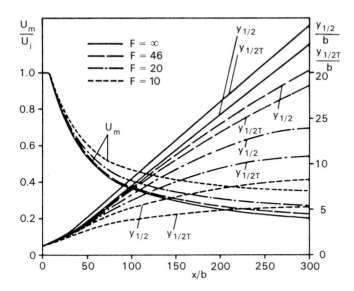

Fig. 5   Variation of maximum velocity U_m, velocity half width $y_{1/2}$ and temperature half width $y_{1/2T}$ in stably stratified wall jets in stagnant surroundings, from Ljuboja and Rodi (1981)

3.4 Vertical Mixing in Stratified Channel Flow

The influence of stable stratification on the vertical mixing due to bed-generated turbulence in open channel flow is considered next. Experiments of Schiller and Sayre (1975) were simulated in which heated water was discharged coaxially at the surface into developed channel flow. The flow situation is sketched in Fig. 6. Discharge and channel velocities were the same so that very little turbulence was generated by the discharge and virtually all the turbulence originated from generation near the channel bed. Schiller and Sayre studied mainly the influence of buoyancy on the development of the temperature profiles downstream of the heated water discharge, and especially the tendency towards complete mixing when the profiles become uniform over the depth. The influence of buoyancy is characterised by the exit densimetric Froude number F_0 defined in Fig. 6. At higher discharge Froude numbers, the channel turbulence causes the heated water to mix with the

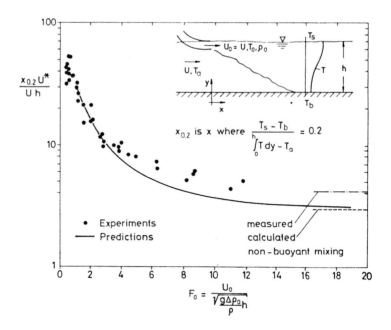

Fig. 6   Vertical mixing in stratified channel flow, from
         Hossain and Rodi (1980)

channel water until the temperature profile is vertically
uniform. With increasing density difference between discharge
and channel water (decreasing Froude number), the mixing is
reduced and eventually ceases entirely when the Froude number
is close to unity. This influence is shown in Fig. 6, where
the mixing distance $x_{0.2}$ defined in the figure characterises

the rate of mixing, i.e. the rate at which vertically uniform
temperature profiles are approached.

    Hossain (1980) tested the model first for the case of
unstratified channel flow in order to ensure that it yielded
realistic results for the background flow. Good agreement

was obtained for the vertical variation of U, $\overline{v^2}$, k, ε and the
eddy viscosity v_t with the experimental data of Nakagawa et al.

(1975) and Ueda et al. (1977). The calculation of the Schiller
and Sayre experiment was started by adjusting the roughness
coefficient E in the logarithmic law of the wall (23) to obtain
the velocity profile measured by Schiller and Sayre in their
channel without heated water discharge. As the experiment was

carried out with a roughened channel bed, a relatively low
value of E = 0.2 had to be used. The calculations for the
stratified situations were carried out by first calculating
unstratified channel flow to development and using the results
as initial profiles for the ambient flow. As there is great
uncertainty about the details of the initial conditions for
the heated water discharge, uniform distributions of the discharge were assumed with low turbulence values at the outlet.

Schiller and Sayre have found that, when they non-
dimensionalised the mixing distance $x_{0.2}$ with the channel
depth h and the ratio of average velocity $\bar{U}$ to friction velocity U_*, this distance correlates quite well with the exit
Froude number F_0. Fig. 6 shows this correlation together with
the predicted results. There is a large increase in the
mixing distance when the Froude number falls below 10, and this
distance tends to infinity when the Froude number approaches
unity, indicating that the discharged warm water does not mix
at all with the channel water and a two-layer situation
develops. The model can be seen to simulate this behaviour
very well. The reader is reminded that the surface correction
was used which damps the vertical fluctuations. Without this
correction, the mixing distance $x_{0.2}$ would have been too large
by a factor of 2 for the non-buoyant case. The correction
reduces the vertical fluctuations near the surface and thus
the initial mixing, but the predicted non-buoyant mixing is
still somewhat larger than the measured one, which is due to
the fact that in the non-buoyant case the turbulence model
implies a turbulent Prandtl number of 0.8 while measurements
indicate that σ_t is about unity.

3.5 *Plane Wake in Stratified Environment*

The last example concerns the simulation of an experiment
by Monroe and Mei (1968) who pulled a circular cylinder
through water with a linear stable stratification in a towing
tank. Close to the cylinder, a wake developed like in homogeneous medium, but beyond 10 cylinder diameters the influence
of the stable stratification started to damp the turbulence
and to reduce the wake spreading. This influence continued
until the wake ceased to grow, the turbulence collapsed and
internal waves were formed. This formation of internal waves
cannot be simulated with the present model; it could only be
reproduced if a suitable time-dependent calculation scheme
were employed.

Hossain (1980) calculated Monroe and Mei's experiments both for the wake in unstratified surroundings and for the wake in stably stratified environment characterised by the densimetric Froude number F_N = 98 as defined in Fig. 7. The calculations were started at x/D = 20 with the measured width of the wake and with profiles of velocity, k and ε corresponding to developed plane wake flow (for details see Hossain, 1980). The calculated development downstream of this station is compared with the measurements in Fig. 7 for both the wake in neutral and stratified environment. In neutral environment the half-width of the wake grows with $x^{1/2}$ as expected from similarity analysis, and there is good agreement with Schlichting's (1968) empirical relation. The reduction of growth of the wake by the stratification and the cessation of this growth are also simulated very well by the model.

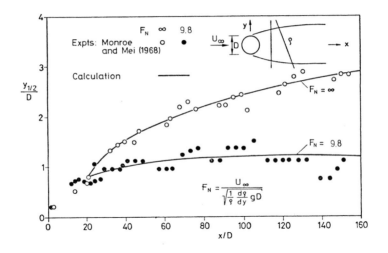

Fig. 7   Plane wake in stably stratified environment, growth of half width $y_{1/2}$, from Hossain (1980)

4.   CONCLUDING REMARKS

In this paper, a buoyancy-extended version of the widely used k-ε model was presented and discussed. In addition to the buoyancy production or destruction terms appearing in the k- and ε-transport equations of the basic version of the model under buoyant situations, the constants c_μ and σ_t (turbulent

Prandtl number) were replaced by buoyancy-dependent functions. These were derived by simplifying the differential transport equations for the individual turbulent stresses and heat fluxes of a particular second-order closure model to algebraic expressions for these stresses and fluxes. The second-order closure model used as starting point also accounts for the effect of the damping of the lateral fluctuations by the presence of a solid or free surface. In the process of simplifying the differential equations, the buoyancy and surface-damping terms appearing as source terms were fully retained. Hence the resulting c_μ and σ_t functions account for buoyancy and surface-damping effects in the same way as do the original transport equations.

The buoyancy-extended k-ε model was tested for a number of horizontal shear layers with stable stratification; mostly discharge situations of interest to pollution problems were considered, but a wake in stably stratified environment was also among the application examples. In all cases, the reduction of mixing and of the spreading of the shear layer due to the stable stratification was simulated well by the model, using the same empirical constants in all cases. One empirical constant multiplying the buoyancy production in the ε-equation had been adjusted by reference to the heated-surface-jet data, but for the horizontal shear layers considered here the results are not very sensitive to the value of this constant. In fact, nearly the same results can be obtained by neglecting altogether the buoyancy term in the ε-equation. This statement is valid only for stably stratified horizontal free shear layers; in unstably stratified layers, near-wall layers and vertical buoyant jets, the buoyancy production/destruction term in the ε-equation is essential, as was discussed above.

Certain details of the flows considered were not reproduced realistically by the model, like the linear vertical temperature distribution observed in buoyant surface jets. Also, for unstably stratified layers as well as for stratified wall layers with heat flux to or from the wall, the ε-equation in its present form is not satisfactory; in the latter it does not reproduce the influence of stratification on the length-scale according to the Monin-Obhoukhov relation. Further research is necessary to develop a more generally applicable form of the ε-equation. As the model presented in this paper is of the eddy viscosity/diffusivity type, it is not suitable for certain geophysical flow situations in which the heat flux is against the temperature gradient (counter-gradient flux); such cases can only be simulated with a full second-order closure scheme. However, there are many situations, especially associated with discharges into the environment, where the simpler model presented here can give answers of sufficient accuracy for practical purposes.

5. ACKNOWLEDGEMENTS

The work reported here was supported by the Deutsche Forschungsgemeinschaft via the Sonderforschungsbereich 80. The author is grateful to Mrs. R. Zschernitz for the careful preparation of the manuscript.

6. REFERENCES

Betts, P.L. and Haroutunian, V. (1983 A k-ε Finite Element Simulation of Buoyancy Effects in the Atmospheric Surface Layer, ASME paper 83-WA/HT-32.

Celik, I., Hossain, M.S. and Rodi, W. (1982) Modelling of Free-Surface Proximity Effects on Turbulence. Proceedings International Symposium on Refined Modelling of Flows, Paris.

Ellison, T.H. and Turner, J.S. (1959) Turbulence Entrainment in Stratified Flow. *J. Fluid Mech.*, **6**, pp. 423-488.

Gibson, M.M. (1983) Private communication.

Gibson, M.M. and Launder, B.E. (1976) On the Calculation of Horizontal Non-Equilibrium Turbulent Flows under Gravitational Influence. *J. Heat Transfer,* ASME, **98**, Cl, pp. 81-87.

Gibson, M.M. and Launder, B.E. (1978) Ground Effects on Pressure Fluctuations in the Atmospheric Boundary Layer. *J. Fluid Mech.*, **86**, pp. 491-511.

Goussebaile, P. and Viollet, P.L. (1982) On the Modelling of Turbulent Flows under Strong Buoyancy Effects in Cavities with Curved Boundaries. Proceedings International Symposium on Refined Modelling of Flows, Paris.

Hossain, M.S. (1980) Mathematische Modellierung von turbulenten Auftriebsströmungen. Ph.D. thesis, University of Karlsruhe.

Hossain, M.S. and Rodi, W. (1980) Mathematical Modelling of Vertical Mixing in Stratified Channel Flow. Proceedings 2nd International Symposium on Stratified Flows, Trondheim, Norway.

Hossain, M.S. and Rodi, W. (1982) A Turbulence Model for Buoyant Flows and its Application to Vertical Buoyant Jets. In "Turbulent Buoyant Jets and Plumes", (W. Rodi, ed.), HMT-Series, Vol. 6, Pergamon Press, Oxford, England.

Launder, B.E. and Spalding, D.B. (1974) Numerical Computation of Turbulent Flows, *Computer Meth. in Appl. Mech. and Engg.*, **3**, pp. 269-289.

Leschziner, M.A. (1979) Numerical Prediction of the Internal Density Jump, Proceedings 18th IAHR Congress, Cagliari, Italy.

Ljuboja, M. and Rodi, W. (1980) Calculation of Turbulent Wall Jets with an Algebraic Reynolds Stress Model. *ASME J. of Fluid Engg.*, **102**, pp. 350-356.

Ljuboja, M. and Rodi, W. (1981) Prediction of Horizontal and Vertical Turbulent Buoyant Wall Jets, *ASME J. of Heat Transfer*, **103**, pp. 343-349.

McGuirk, J.J. and Rodi, W. (1979) Mathematical Modelling of Three-Dimensional Heated Surface Jets, *J. Fluid Mech.*, **95**, pp. 609-633.

Monroe, R.H., Jr. and Mei, C.C. (1968) The Shape of Two-Dimensional Wakes in Density-Stratified Fluids. MIT Hydrodynamics Lab., Rept. No. 110.

Nakagawa, H., Nezu, I. and Ueda, H. (1975) Turbulence in Open Channel Flow over Smooth and Rough Beds, Proceedings JIHR 241, pp. 155-168.

Patankar, S.V. and Spalding, D.B. (1970) Heat and Mass Transfer in Boundary Layers, Intertext Book, London, 2nd edition.

Rodi, W. (1979) Influence of Buoyancy and Rotation on Equations for the Turbulent Length-Scale, Proceedings 2nd Symposium on Turbulent Shear Flows, London.

Rodi, W. (1980) Turbulence Models and Their Application in Hydraulics. Book publication of International Association for Hydraulic Research, Delft, The Netherlands.

Rodi, W. (1984) Examples of Turbulence-Model Applications. In "Turbulence Models and their Applications", Collection de la Direction des Études et Recherches d'Electricité de France, Vol. 56, Editions Eyrolles, Paris.

Schlichting, H. (1968) Boundary Layer Theory, 6th Ed.,
    McGraw Hill.

Svensson, U. (1980)   On the Numerical Prediction of Vertical
    Turbulent Exchange in Stratified Flows.   Proceedings 2nd
    International Symposium on Stratified Flows, Trondheim,
    Norway.

Ueda, H., Möller, R., Komori, S. and Mitsushina, T. (1977)
    Eddy Diffusivity Near the Free Surface of Open Channel Flow,
    Int. J. Heat Mass Transfer, **20**, 11, pp. 1127-1136.

Vanvari, M.R. and Chu, V.H. (1974)   Two-Dimensional Turbulent
    Surface Jets of Low Richardson Number.   Fluid Mechanics
    Lab., Tech. Rept. No. 74-2 (FMR), Dept. of Civil Engg. and
    Appl. Mech., McGill University, Montreal, Canada (see also
    Chu, V.H. and Vanvari, M.R., 1976.   Experimental Study
    of Turbulent Stratified Shearing Flow, *ASCE J. Hydraulics
    Div.,* **102**, HY6, pp. 691-706).

UNDERSTANDING AND MODELLING TURBULENCE IN STABLY STRATIFIED
FLOWS BY CONSIDERING DISPLACEMENTS AND MIXING OF FLUID ELEMENTS

J.C.R. Hunt
(University of Cambridge)

ABSTRACT

   The main qualitative effects of imposing a stable stratification on turbulent flows are that

1)   some of the kinetic energy of the turbulence is transformed
     into potential energy and into the energy of internal
     gravity waves;

2)   for given turbulence energy, vertical displacements of
     fluid elements and the vertical flux of matter, heat and
     momentum are reduced;

3)   for given turbulence in equilibrium with a velocity
     gradient, proportionately larger increases in the velocity
     gradient are associated with small changes in stratifi-
     cation;

4)   in steady shear flows the scales of vertical components of
     turbulence decrease and, for given turbulent energy, the
     rate of dissipation increases; but this is not necessarily
     so in homogeneous turbulence.

   The main physical ideas or models that have guided theoretical and experimental work in stably stratified turbulence
have been (a) the potential energy limitations on vertical
motion of parcels of fluid (e.g. Prandtl 1952, p. 391),
(b) the generation and radiation of internal waves at the large
scales (Stewart 1969), and (c) the breaking of waves or billows
and increased dissipation at the smaller scales (e.g. Sherman,
Imberger and Corcos (1978)).

    To explain all the main effects listed above, two additional
concepts seem to be necessary. (d) If there is no mixing of
density between fluid elements, then, because of the finite
energy in the turbulence, the vertical displacements of fluid
elements must be limited, and no vertical diffusion can occur.
Thus it is only because of small-scale mixing that vertical
diffusion is possible. Such mixing enables all scales of
motion to contribute to vertical diffusion, even large-scale
wave motions, and (e) that if the time scale of the mean
velocity gradient is smaller than that of the buoyancy forces,
then the velocity gradients and the turbulent energy largely
determines the small-scale turbulence structure. Near a
rigid surface or strong density interface, the distance to the
surface or interface also has to be considered.

    The mathematical modelling and the physical consequences of
these concepts (Csanady 1964; Pearson, Puttock and Hunt 1983;
Hunt 1982) will be described and compared with turbulence
measurements in the laboratory and the atmospheric boundary
layer (Hunt, Kaimal, Gaynor and Korrel 1982; Britter, Hunt,
Marsh and Snyder 1983; Pearson and Linden 1983), and with the
results of direct computations (Riley, Metcalfe and Weissman
1981) and second-order closure models (Brost and Wyngaard
1978; Gartrell and Pearson 1983; Gibson and Launder 1978).

REFERENCES

Britter, R.E., Hunt, J.C.R., Marsh, L., Snyder, W.H. (1983)
    The effects of stable stratification on turbulent diffusion
    and the decay of turbulence. *J. Fluid Mech.*, **127**, 27-44.

Brost, R.A. and Wyngaard, J.C. (1978) A model study of the
    stably stratified planetary boundary layer. *J. Atmos. Sci.*,
    35, 1428.

Csanady, G.T. (1964) Turbulent diffusion in a stratified
    fluid. *J. Atmos. Sci.*, **21**, 439-447.

Gartrell, G. and Pearson, H.J. (1981) On the scaling of the
    vertical diffusivity in stably stratified flows. Report
    of Keck Lab., Cal. Inst. of Tech., 1981.

Gibson, M.M. and Launder, B.E. (1978) Ground effects on
    pressure fluctuations in the atmospheric boundary layer.
    J. Fluid Mech., **86**, 491-511.

Hunt, J.C.R. (1982) Diffusion in the stable boundary layer.
    In: Atmospheric Turbulence and Air Pollution Modelling,
    (F. T.M. Nieuwstadt and H. van Dop, Eds.), Reidel, 231-274.

Hunt, J.C.R., Kaimal, J.C., Gaynor, J. and Korrell, A. (1983) Observations of turbulence structure in stable conditions at the Boulder Atmospheric Observatory. B.A.O. Report 4, Studies of the Stable Atmosphere at the B.A.O. NOAA Env. Res. Lab. Report, Boulder, Colo. (To appear in *Quat. J. Roy. Met. Soc.*, July 1985).

Linden, P. (1979) Mixing in stratified fluids. *Geo. and Astro. physical fluid dyn.*, **13**, 3, 1979.

Pearson, H.J. and Linden, P. (1983) The final stage of decay of turbulence in stably stratified fluid. *J. Fluid Mech.*, **134**, 195-203.

Pearson, H.J., Puttock, J.S. and Hunt, J.C.R. (1983) A statistical model of fluid element motions and vertical diffusion in a homogeneous turbulent flow. *J. Fluid Mech.* (in press).

Prandtl, L. (1952) The Essentials of Fluid Dynamics. Blackie.

Riley, J.J., Metcalfe, R.W. and Weissman, M.A. (1981) Direct numerical simulations of homogeneous turbulence in density stratified fluids. In: Nonlinear Properties of Internal Waves. La Jolla Inst., Conference Proc. Am. Institute of Physics, No. 76, New York.

Sherman, F.S., Imberger, J. and Corcos, G.M. (1978) Turbulence and mixing in stably stratified waters. *Ann. Rev. Fluid Mech.*, **10**, 267.

Stewart, R.W. (1969) Turbulence and waves in a stratified atmosphere. *Radio Sci.*, **4**, 1269-1278.

THE TURBULENCE MODELLING OF VARIABLE DENSITY FLOWS - A MIXED-WEIGHTED DECOMPOSITION

B.E. Launder

(UMIST)

ABSTRACT

A new decomposition is presented of a variable density turbulent field into mean and fluctuating parts which adopts both the mass-weighted and the conventional mean velocity as dependent variables. By so doing, the different roles played by the velocity vector in the equations of motion - that of fluid transporter and of momentum per unit mass - can be properly accommodated. For the mean-field conservation equations of momentum and species, the effects of turbulence in each equation are confined to a single correlation between fluctuating quantities; indeed, the form is identical to that for a uniform density flow. The transport equations for the turbulent correlations are more complex than when mass-weighted averages are adopted but, arguably, the second-moment closure methodology is clearer to apply. The modelling of the transport processes involving density fluctuations is discussed in connection with the behaviour observed by Rebollo in the variable density mixing layer. The effective turbulent Prandtl-Schmidt number in a variable density medium is shown to be crucially sensitive to the specific forms of the approximation for correlations containing the fluctuating pressures.

REFERENCES

Ha Minh, H., Launder, B.E. and MacInnes, J.M. (1982) "The turbulence modelling of variable density flows - a mixed-weighted decomposition", Turbulent Shear Flows-3 (Eds. L.J.S. Bradbury, et al), 291-308, Springer Verlag, Berlin.

MacInnes, J.M. (1985) Turbulence modelling of flows with non-uniform density, PhD Thesis, Faculty of Technology, University of Manchester.

2. TURBULENT STRATIFIED FLOWS IN THE ENVIRONMENT

A MODEL FOR THE STATIONARY, STABLE BOUNDARY LAYER

F.T.M. Nieuwstadt

(Royal Netherlands Meteorological Institute)

ABSTRACT

Based on the hypothesis that the Richardson number, Ri, and the flux Richardson number, Ri_f, are constant with height, we are able to solve the equations for the steady-state, stable boundary layer in closed form. We find that the height of the boundary layer is given by the well-known equation proposed by Zilitinkevich (1972): $h = c(u_* L/f)^{\frac{1}{2}}$, where u_* is the friction velocity, L the Obukhov length and f the Coriolis parameter. For the constant c we obtain $c^2 = \sqrt{3} \, k \, Ri_f$, where k is the Von Karman constant. Profiles for mean and turbulence quantities are explicitly calculated. For instance, the magnitude of the turbulent stress is given by $\|\tau\|/u_*^2 = (1-z/h)^{3/2}$. These results provide in many respects an acceptable description of the observed nocturnal boundary layer. The model fails near the top of the boundary layer, where the solution exhibits a singular behaviour.

1. INTRODUCTION

A stable boundary layer is often encountered during night-time over a land surface. It frequently takes the form of a shallow turbulent layer. The purpose of this paper is to study a model for the nocturnal boundary layer.

The dynamics of this boundary layer may be dominated by more than one physical process. Apart from turbulence, long-wave radiation and gravity waves can also play an important role. Recent progress with respect to the influence of long wave radiation is reported by Garratt and Brost (1981) and André and Mahrt (1982). Einaudi and Finnigan (1981) and Finnigan and Einaudi (1981) discuss the importance of gravity waves. Here, we focus only on a stable boundary layer which is dominated by turbulent processes. We realize that a complete model for the

nocturnal boundary layer must necessarily encompass all
aspects. Such a task can only be attempted, when the influence
of all individual processes is thoroughly understood.

Previous modelling studies have become increasingly complex.
They evolved from simple K-models (Delage, 1974) to a full
second-order closure technique (Wyngaard, 1975; Brost and
Wyngaard, 1978). The results of these models have greatly
increased our insight in the processes that govern the stable
boundary layer. In this study we return to a more simple
description of turbulence. It is based on a local scaling
hypothesis for turbulence in stable conditions (Nieuwstadt,
1984). With the aid of this turbulence model we aim to study
the vertical structure of the stable boundary layer.

The type of stable boundary layer that we consider here,
consists of a layer of continuous turbulence adjacent to the
surface. No intermittency or patchiness is taken into account.
The height of the turbulent layer is defined as the boundary-
layer height, h. The main characteristic of this stable
boundary layer is that turbulence supports a negative heat flux
throughout its depth.

In the course of this study we will compare the results of
our model with observations which were taken on the meteorological mast at Cabauw. An extensive discussion of these
measurements is given by Nieuwstadt (1984). Therefore, a short
summary will suffice here. During seven clear nights we
measured variables such as wind velocity and temperature along
the mast between 20 and 200 m. The data were sampled with a
frequency of 5 Hz and were high-pass filtered with a cut-off
frequency of 0.01 Hz. From these filtered time series we
calculated the turbulent variances and covariances over a time
period of 30 minutes. Mean wind and temperature observations
are available over concurrent time periods. The height of the
boundary layer was observed with a monostatic acoustic sounder.

To conclude we give an outline of the paper. In section 2
we present the turbulence model. It furnishes the closure
hypothesis for the equations of the stable boundary layer which
we derive in section 3. The solution of these equations is
then discussed in section 4.

2. TURBULENCE MODEL

In this section a closure hypothesis is formulated for turbulence in stable conditions. The basis of our approach is the
assumption that turbulence obeys local scaling in the stable
boundary layer, where the term local denotes values measured
at the same height. Local scaling is extensively discussed by

Nieuwstadt (1984). There, we have found that dimensionless combinations of variables observed at the same height approach a constant value at a sufficient distance from the surface. Examples of these dimensionless, local variables are the Richardson number, Ri, and the flux Richardson number, Ri_f. Local scaling thus implies

$$Ri \equiv \frac{g}{T} \frac{\partial \theta}{\partial z} \Big/ \left\| \frac{\partial \underset{\sim}{U}}{\partial z} \right\|^2 = const, \tag{1a}$$

$$Ri_f \equiv - \frac{g}{T} \overline{w\theta} \Big/ \left(\underset{\sim}{\tau} \cdot \frac{\partial \underset{\sim}{U}}{\partial z} \right) = const, \tag{1b}$$

where $\partial \underset{\sim}{U}/\partial z = (\partial U/\partial z, \partial V/\partial z)$ and $\partial \theta/\partial z$ are the mean velocity and temperature gradient; $\underset{\sim}{\tau} = (-\overline{uw}, -\overline{vw})$, and $\overline{w\theta}$ are the momentum and temperature flux (note that to simplify notation we have omitted the density from the definition of momentum flux). We now propose (1a) and (1b) as a turbulence parameterization for the stable boundary layer.

In the remaining part of this section we will summarize some arguments in support of the result (1a) and (1b). The analysis starts with the simplified equations for the turbulent kinetic energy and the temperature variance

$$\underset{\sim}{\tau} \cdot \frac{\partial \underset{\sim}{U}}{\partial z} + \frac{g}{T} \overline{w\theta} - \varepsilon = 0, \tag{2a}$$

$$-\overline{w\theta} \frac{\partial \theta}{\partial z} - \varepsilon_\theta = 0, \tag{2b}$$

where ε and ε_θ denote the molecular destruction terms. These equations are obtained from the complete turbulent kinetic energy and temperature variance budgets (Businger, 1982) by neglecting the time variation, advection and flux divergence terms. The first two terms can be neglected because turbulence is taken to be quasi-stationary and horizontally homogeneous. Omission of the flux divergence term is justified by the assumption that vertical transport is small in stable conditions (Brost and Wyngaard, 1978).

A solution of (2a) and (2b) requires estimates for ε and ε_θ. A fundamental hypothesis of turbulence theory states that

$\varepsilon \sim u^3/\ell$ and $\varepsilon_\theta \sim u\,\theta^2/\ell$, where u, θ are respectively a fluctuating velocity and temperature and where ℓ is a length scale
characteristic of the macroscopic structure of turbulence
(Tennekes and Lumley, 1972). In the spirit of local scaling,
u and θ are taken to be proportional to the local momentum and
temperature flux. The result reads

$$\varepsilon \simeq \|\underset{\sim}{\tau}\|^{3/2}/\ell, \tag{3a}$$

$$\varepsilon_\theta \simeq (\overline{w\theta})^2 / (\ell\,\|\underset{\sim}{\tau}\|^{\frac{1}{2}}). \tag{3b}$$

To complete our analysis we need an expression for ℓ.
Taking a balance between inertia forces, $\sim\|\underset{\sim}{\tau}\|/\ell$, and buoyancy
forces, $\sim g/T\,\ell\,\partial\theta/\partial z$, Brost and Wyngaard (1978) propose

$$\ell \simeq \|\underset{\sim}{\tau}\|^{\frac{1}{2}}/N, \tag{4}$$

where N is the Brunt-Vaisala frequency equal to
$N = (g/T\,\partial\theta/\partial z)^{\frac{1}{2}}$. The proportionality constant in (4) is
estimated in the surface layer to be ~ 0.5 (Hunt, 1982).

With the aid of (3a), (3b) and (4) we can solve equations
(2a) and (2b). It can be easily checked that the result leads
to (1a) and (1b). The constant Richardson and flux Richardson
number are thus consistent with the energy and temperature
variance budgets. At this stage we may mention that this
conclusion does not depend on the particular choice (4) for
the length scale ℓ. Any expression of the form
$\ell = f\ (\|\underset{\sim}{\tau}\|^{\frac{1}{2}},\ \partial U/\partial z,\ g/T\,\partial\theta/\partial z)$ will lead to the same result.

The justification of (1a) and (1b) is now evident. Everywhere in this flow turbulence must satisfy a local equilibrium
between production and destruction processes. This equilibrium
is not a function of the location in the boundary layer and
consequently the Ri and Ri_f number become equal to a constant
independent of height. This quality of the stable boundary
layer is also denoted as "z-less" stratification (Wyngaard,
1973).

Let us next review some observational evidence. The measurements of Mahrt (1979; Figure 4) and of Garratt (1982; Figure 2)
support a Richardson number which is independent of height.
Our observations are shown in Fig. 1. Each data point in this
figure gives the average of all observations taken within the

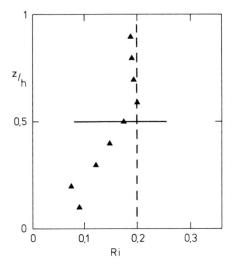

Fig. 1   The Richardson number as a function of non-dimensional
         height. Each point indicates the average of all obser-
         vations within the same height range. The horizontal
         bar indicates the standard deviation of the data.

same height range. The variation between the individual experiments is also indicated. From this figure we may conclude that
on the average Ri = const. is a reasonable assumption for the
upper part of the boundary layer. For the value of the constant
we will adopt 0.2. The same value is taken for the flux
Richardson number.

   The measurements also indicate that (1a) and (1b) break down
close to the surface. This is in agreement with the well-known
result of surface-layer turbulence, according to which Ri and
Ri_f approach zero for $z \to 0$. The reason for this behaviour is
the fact that the length scale ℓ becomes proportional to z when
$z \to 0$, a property which was not included in our expression (4).
However, as a first approach we shall neglect this variation of
Ri and Ri_f near z = 0 in the forthcoming analysis. A more
detailed investigation of the solution close to the surface is
postponed to Appendix A. There we derive that our results for
constant Ri and Ri_f can be considered as the first term in an
asymptotic series for $h/L \gg 1$, where L is the Obukhov length
defined in (8c).

To conclude this section we summarize our hypothesis concerning the structure of the stable boundary layer. We have seen that for $h/L \gg 1$ the majority of the boundary layer is governed by a local equilibrium so that Ri and Ri_f become constant. An alternative explanation of this result is that in this region turbulent eddies do not feel the presence of the ground because the turbulence length scale, ℓ, is smaller than z. As a result the boundary layer behaves like a free shear layer in a stratified flow, a case for which a constant Ri number is confirmed. In the next sections we analyze the solution of such a free shear layer which matches to the boundary conditions at $z = 0$. A departure from this picture may be expected in the surface layer where the assumption that $\ell < z$ no longer holds. Furthermore, we may expect deviations near the top of the boundary layer because in that region the Ri must jump to a value above the critical Richardson number.

3.   EQUATIONS

In this section we develop a non-dimensional set of equations for the stable boundary layer. We depart from the well-known equations that describe the evolution of the mean temperature and the mean velocity profile in a horizontally homogeneous boundary layer. They read

$$\frac{\partial \Theta}{\partial t} = - \frac{\partial \overline{w\theta}}{\partial z} , \tag{5a}$$

$$\frac{\partial W}{\partial t} = - if (W - W_g) + \frac{\partial \tau}{\partial z} , \tag{5b}$$

$$\frac{g}{t} \frac{\partial \Theta}{\partial z} / \left\| \frac{\partial W}{\partial z} \right\|^2 \equiv Ri = const, \tag{5c}$$

$$- \frac{g}{T} \overline{w\theta} / (\tau^* \frac{\partial W}{\partial z}) \equiv Ri_f = const, \tag{5d}$$

where Θ is the mean temperature. Furthermore we use a complex notation in (5b), so that $W = U + iV$ is the mean velocity, $W_g = U_g + iV_g$ is the geostrophic wind and $\tau = \tau_x + i \tau_y$ is the turbulent stress. (A star indicates a complex conjugate.) To complete the set of equations we have added the closure hypotheses (1a) and (1b). Note that in (5d) we have taken the stress to be aligned with the velocity gradient.

In this study we shall restrict ourselves to the case of a steady-state, stable boundary layer. By definition this is a boundary layer, in which turbulent fluxes do not depend on time. Nocturnal boundary layers are rarely stationary, but they evolve to a steady-state as an asymptotic limit for large times (Nieuwstadt and Tennekes, 1981). By means of a comparison with observations we will see in the following whether this solution of stationary boundary layer is representative for the structure of the real nocturnal boundary layer.

We have introduced the stationary boundary layer as being characterized by fluxes that are independent of time. This leads to the following consequences. First we find from (5a) that the cooling rate, $d\theta/dt$, must be constant or the temperature decreases linearly with time. Furthermore (5d) shows that the velocity gradient must be independent of time and consequently it follows from (5c) that the temperature gradient is also independent of time. It is, therefore, advantageous to consider equations for these gradients, because time derivatives will then disappear. Such a set of equations is easily found from (5a) - (5d). It reads

$$O = \frac{\partial^2 \overline{w\theta}}{\partial z^2} \,, \tag{6a}$$

$$O = - \text{ if } \frac{\partial W}{\partial z} + \frac{\partial^2 \tau}{\partial z^2} \,, \tag{6b}$$

$$\frac{g}{T} \frac{\partial \theta}{\partial z} \Big/ \left\| \frac{\partial W}{\partial z} \right\|^2 \equiv Ri = \text{const,} \tag{6c}$$

$$- \frac{g}{T} \overline{w\theta} \Big/ (\tau^* \frac{\partial W}{\partial z}) \equiv Rif = \text{const.} \tag{6d}$$

Here, we have also assumed a barotropic atmosphere, so that $\partial W_g/\partial z = O$.

Before we attempt to solve this set of equations, we can already obtain some information by an appropriate scaling of the individual terms. For that reason we propose the following non-dimensional parameters.

$$\sigma = \tau/u_*^2, \tag{7a}$$

$$H = \overline{w\theta}/\overline{w\theta}_o, \tag{7b}$$

$$\eta = z/h, \tag{7c}$$

$$s = \frac{L}{u_*} \frac{\partial W}{\partial z}, \tag{7d}$$

$$\gamma = \frac{L}{T_*} \frac{\partial \theta}{\partial z}, \tag{7e}$$

where the friction velocity u_*, the surface heat flux $\overline{w\theta}_o$, the temperature scale T_* and the Obukhov-length L are defined by

$$u_*^2 = \| \tau(0) \|, \quad \overline{w\theta}_o = w\theta(0), \tag{8a}$$

$$T_* = - \overline{w\theta}_o/u_*, \tag{8b}$$

$$L = - u_*^3/(k \ g/T \ \overline{w\theta}_o), \tag{8c}$$

where k is the Von Karman constant.

Equations (7a) - (7c) follow the rather straightforward procedure to scale velocity in terms of u_*, temperature in terms of T_* and height in terms of h (Caughey et al., 1979). The exception is the scaling of the temperature and velocity gradients (7d) and (7e), where we use L instead of h. It expresses the fact that from observation we expect gradients to be large in a stable boundary layer (remember that we look at cases for which h/L >> 1). An additional argument in favour of this scaling is that it leads to results consistent with the well-established Monin-Obukhov similarity theory in the surface layer, as we shall see below.

Substitution of the scaled variables in (6a) - (6d) leads to our final set of equations for the steady-state, stable boundary layer

$$\frac{\partial^2 H}{\partial \eta^2} = 0 \tag{9a}$$

$$\frac{\partial^2 \sigma}{\partial \eta^2} - i\, c^2 s = 0, \tag{9b}$$

$$\frac{\gamma}{\|s\|^2} = k\ Ri, \tag{9c}$$

$$\frac{H}{\sigma^* s} = k\ Ri_f. \tag{9d}$$

In these equations appears a non-dimensional parameter c. It is defined by

$$c = h \left/ \left[\frac{u_* L}{f}\right]^{\frac{1}{2}} \right. . \tag{10}$$

This expression can be alternatively interpreted as an equation for the height of the stationary stable boundary layer. As a matter of fact, it is identical to the relation proposed by Zilitinkevich (1972). Our scaling approach is thus consistent with this well-known result. Moreover, we find from (9b) that the constant c in Zilitinkevich's equation is given by the ratio between the turbulent stress and Coriolis term in the momentum equation. Its numerical value, however, can only follow from an explicit solution of (9a) - (9d).

To obtain such a solution we must specify boundary conditions. We propose

$$\sigma = 1,\ H = 1 \text{ for } \eta = 0 \tag{11a}$$

$$\sigma = 0,\ H = 0 \text{ for } \eta = 1. \tag{11b}$$

The first condition expresses that the turbulent fluxes must approach their surface values (8a) at $\eta = 0$. The second condition states that turbulence vanishes at $z = h$.

The equations (9a) - (9d), together with the boundary conditions (11a) and (11b), form a closed set. It is the starting point for our discussion of the next section.

4. RESULTS

We show in this section that the set of equations (9a) - (9d) with the boundary conditions (11a) - (11b) allow an explicit solution. The discussion is divided into four subsections. In the first we derive the solution for the turbulent fluxes. The mean velocity and temperature profiles are discussed in the second subsection. In the third we shall consider profiles of turbulence parameters such as variances, dissipation rates and structure parameters. Finally we shall examine the behaviour of the solution close to the top of the boundary layer.

(a) Fluxes

The first part of our solution follows immediately from (9a), which leads to following temperature flux profile

$$\overline{w\theta}/\overline{w\theta}_O = 1 - z/h. \tag{12}$$

Our observations, shown in Fig. 2, seem to agree with such a linear profile. The same result is reported in Mahrt et al. (1979; Fig. 4). On the other hand Caughey et al. (1979) find a temperature flux profile which decreases faster with height than the linear profile (12). Their measurements were taken in the early evening when conditions are probably far from steady state. However, in a recent study Garratt (1981) reanalyzed these data and claims that they can also be fitted to a linear profile. Therefore, we conclude that a linear heat flux profile seems realistic for the stable boundary layer.

Next we consider the turbulent stress. With (12) we can eliminate s from (9a) and (9d) to yield an equation for σ only, which reads

$$\sigma * \frac{\partial^2 \sigma}{\partial \eta^2} - i \, c^2 \, \frac{1}{k \, Ri_f} \, (1 - \eta) = 0. \tag{13}$$

A solution of this equation which satisfies the boundary conditions (11a) and (11b) is given by

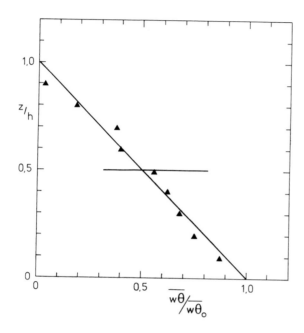

Fig. 2 The temperature flux $\overline{w\theta}$, non-dimensionalized with its surface value, as a function of non-dimensional height z/h. For an explanation of the symbols see Fig. 1. The solid line is equation $\overline{w\theta}/\overline{w\theta}_o = 1 - z/h$.

$$\tau/u_*^2 = (1-z/h)^{\frac{1}{2}}(3 + i\ \sqrt{3}) \tag{14}$$

To obtain (14) we have to fix the constant c^2 equal to

$$c^2 = \sqrt{3}\ k\ Ri_f \tag{15}$$

We thus find that the constant in Zilitinkevich's equation (10) for the boundary-layer height has a unique value which solely depends on $k\ Ri_f$. With realistic values, $Ri_f \sim 0.2$ and $k \sim 0.4$, we find $c \sim 0.37$, which is close to the observed value ~ 0.4 (Garratt, 1981).

Furthermore, equation (14) shows that the absolute value of the stress follows a 3/2-power law as a function of $(1 - z/h)$. Such a relationship is consistent with observations as shown in

Fig. 3. The data of Caughey et al. (1979) show the same
behaviour. We emphasize that this profile does not depend on
the value of Ri and Ri_f. It is a direct consequence of the

closure hypothesis. Therefore, the agreement with observations
strengthens our confidence in the turbulence model of section
2.

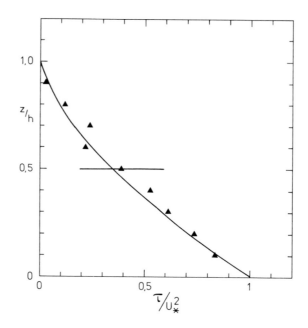

Fig. 3   The absolute value of the momentum flux non-dimensiona-
         lized with its surface value as a function of non-
         dimensional height z/h. For an explanation of the
         symbols see Fig. 1. The solid line is the equation
         $\| \tau \| /u_*^2 = (1 - z/h)^{3/2}$.

(b) Mean profiles

    To derive the mean velocity and temperature profile we first
obtain an expression for the gradients s and γ. They are found
by substitution of (12) and (14) into (9c) and (9d). The
results are

$$s = \frac{1}{k\ Ri_f}\ (1-\eta)^{\frac{1}{2}}\ (-1 + i\ \sqrt{3}), \qquad (16a)$$

$$\gamma = \frac{Ri}{k\ Ri_f^2}\ (1-\eta)^{-1}. \qquad (16b)$$

Let us first consider the velocity profile. Integration of (16a) leads to

$$\frac{W_g - W}{u_*} = \frac{1}{k\ Ri_f}\ \frac{h}{L}\ e^{-i\frac{\pi}{3}}\ (1 - \frac{z}{h})^{\frac{1}{2}(1 + i\ \sqrt{3})}, \qquad (17)$$

where we have used the boundary condition $W = W_g$ at $z = h$. Furthermore, we require that this velocity profile also satisfies $W = 0$ at the surface, $z = 0$. The following relations then result

$$\frac{\|W_g\|}{u_*} = \frac{1}{k\ Ri_f}\ \frac{h}{L}, \qquad (18a)$$

$$\alpha_g = \pi/3, \qquad (18b)$$

where α_g is the angle between the surface wind direction and the geostrophic wind. We recognize (18a) as a resistance law. A similar relationship between $\|W_g\|/u_*$ and h/L was obtained by Brost and Wyngaard (1978), who find a constant of proportionality equal to ~ 11.8. With $k \sim 0.4$ and $Ri_f \sim 0.2$ we find 12.5, an acceptable agreement.

The wind direction change across the boundary layer, α_g, turns out to be 60°. It is independent of stability and of the values for Ri and Ri_f. Again this result closely follows the calculations of Brost and Wyngaard (1978), who find $\alpha_g \sim 58^\circ$ for $h/L \gg 1$.

We note that the velocity profile (17) is a member of the so-called equiangular solutions of the Ekman equations, which were introduced by Lettau and Dabberdt (1970). These solutions

are characterized by a constant angle between the stress and
the velocity defect W_g - W. It is easily checked that in our
case this angle is $\pi/3$.

   Next we consider the asymptotic behaviour of the velocity
profile for $z \to 0$. From (17), (18a) and (18b) follows

$$U/u_* = \frac{1}{k\ \mathrm{Ri}_f}\ z/L,\tag{19a}$$

$$V/u_* = 0.\tag{19b}$$

The result is a linear profile, which is in agreement with
Monin-Obukhov similarity theory for the stable surface layer
(Businger et al., 1971). The profile (19a) is also numerically
consistent with surface-layer observations, if we take
$\mathrm{Ri}_f \sim 0.2$. However, the agreement breaks down for z close to
zero, where the profile should exhibit the well-established
logarithmic form. The cause for this discrepancy is our
assumption that the validity of the closure hypotheses (1a) and
(1b) extends to z = 0. As a matter of fact this assumption
also accounts for the absence of the familiar logarithmic term
in the resistance law (18a). A more detailed treatment of the
solution near z = 0 is given in appendix A, where also an
explicit correction term is calculated.

   Furthermore, it is illuminating to present the velocity profile in another form. With the aid of (18a) we can rewrite
(17) as

$$\frac{W_g - W}{\|W_g\|} = e^{-i\frac{\pi}{3}}\ (1-z/h)^{\frac{1}{2}(1 + i\ \sqrt{3})}.\tag{20}$$

This result shows that the proper scale for the velocity defect
law in a stable boundary layer is the geostrophic wind speed
rather than the friction velocity. The latter is usually taken
as the scaling velocity in a neutral boundary layer (Tennekes,
1982).

   To finish our discussion of the velocity profile we show in
Fig. 4 the hodograph of the velocity and stress profiles.
Furthermore, the wind speed and wind direction profiles are
given in Fig. 5. We see that the wind speed profile is
approximately linear over a large part of the boundary layer.

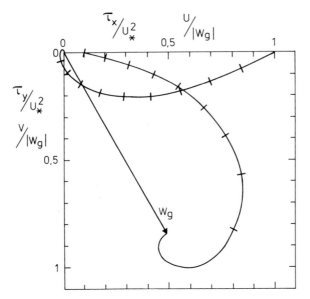

Fig. 4 Hodographs of the wind velocity and the stress profile
given by (20) and (14), respectively. Marks along the
curves indicate values of z/h which are a multiple of
0.1.

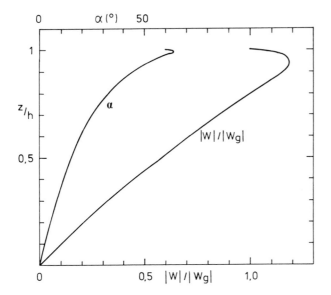

Fig. 5 The wind speed and wind direction profile obtained from
(20) as a function of dimensionless height.

This behaviour is confirmed by observed profiles of the wind
speed in the nocturnal boundary layer. An example of our
observation is for instance shown in Fig. 6. Moreover, the
case studies described by Mahrt et al. (1979) show a similar
picture. We also find that the wind speed becomes supergeo-
stropic in the upper part of the boundary layer and develops
there a so-called low level maximum. This phenomenon is
usually connected to an inertial oscillation at the top of the
boundary layer. Our result shows that it may also occur in
steady state conditions. Fig. 5 also shows that the wind
direction shear is primarily confined to the upper part of the
boundary layer. This behaviour is reasonably well followed by
the observed wind direction profile shown in Fig. 6. However,
the case study of Fig. 6 also illustrates that observations
usually lead to a value of wind direction change across the
boundary layer which is smaller than $\pi/3$.

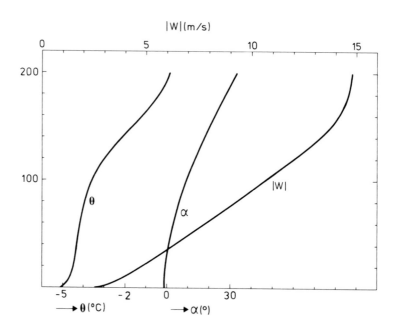

Fig. 6   A case of observed profiles for the wind speed $\|w\|$, the
         wind direction α and the potential temperature θ in a
         stable boundary layer.

We now turn to the mean temperature profile. Integration of (16b) and application of the boundary condition $\Theta = \Theta_o$ at $z = 0$ lead to

$$\frac{\Theta - \Theta_o}{T_\star} = - \frac{Ri}{k \, Ri_f^2} \frac{h}{L} \ln(1 - \frac{z}{h}). \qquad (21)$$

Let us first consider this temperature profile close to the surface. For $z \to 0$ we find

$$\frac{\Theta - \Theta_o}{T_\star} = \frac{1}{k} \frac{Ri}{Ri_f^2} \frac{z}{L}. \qquad (22)$$

This result again satisfies Monin-Obukhov similarity and it matches the linear profile found by Businger et al. (1971) for the stable surface layer. Numerical consistency with observations requires $Ri \sim Ri_f \sim 0.2$.

Above the surface layer the temperature profile is given by a logarithmic function. The curvature ($\sim\partial^2\Theta/\partial z^2$) of this profile is positive. Such behaviour is confirmed by the observed profile shown in Fig. 6. Furthermore, André and Mahrt (1982) state that observed temperature profiles frequently exhibit a positive curvature in a stable boundary layer dominated by turbulence, which agrees with this study. However, they call this case "quasi-mixed". Our analysis shows that a positive curvature is merely a consequence of the turbulence model.

The last subject in our discussion of mean profiles is the exchange coefficients. They are easily derived from the flux profiles (12) and (14) and from the gradients (16a) and (16b). The results are

$$K_M = k \, Ri_f \, u_\star \, L \, (1 - \frac{z}{h})^2, \qquad (23a)$$

$$K_H = k \, \frac{Ri_f^2}{Ri} \, u_\star \, L \, (1 - \frac{z}{h})^2, \qquad (23b)$$

where K_M is the eddy viscosity and K_H is the transfer coefficient for heat.

We note that the exchange coefficients reach their maximum value at $z = 0$. This contradicts surface layer theory which requires $K \sim z$ for $z \to 0$. The reason is again our assumption that (1a) and (1b) are uniformly valid across the boundary layer. In appendix A we shall come back to this problem and introduce a modification of (23a) to incorporate the correct behaviour for $z \to 0$.

The last remark concerns the quadratic dependence of K_M and K_H on $(1-z/h)$, a profile which has been also hypothesized by Misra (1979). In contrast Brost and Wyngaard (1978) propose a $3/2$ - power law based on curve fitting to their model results.

(c)   Turbulent structure

Based on the foregoing results we can now derive the vertical profiles of several turbulence quantities. We start with the production terms in the turbulent energy and temperature variance budget (2a) and (2b). They can be directly calculated from (12), (14), (16a) and (16b). The results are

$$\underset{\sim}{\tau} \cdot \frac{\partial \underset{\sim}{U}}{\partial z} = \frac{1}{k \, Ri_f} \frac{u_*^3}{L} (1 - \frac{z}{h}), \tag{24a}$$

$$- \overline{w\theta} \frac{\partial \Theta}{\partial z} = - \frac{Ri}{k \, Ri_f^2} \frac{T_*}{L} \overline{w\theta}_0 \tag{24b}$$

The budget equations (2a) and (2b) express a balance between the production terms and the molecular destruction term ε and ε_θ. Therefore, profiles of ε and ε_θ can be directly obtained from (24a) and (24b). In dimensionless form they become

$$\frac{kh\varepsilon}{u_*^3} = \frac{1 - Ri_f}{Ri_f} \frac{h}{L} (1 - \frac{z}{h}), \tag{25a}$$

$$\frac{kh\varepsilon_\theta}{T_*^2 u_*} = \frac{Ri}{Ri_f^2} \frac{h}{L}. \tag{25b}$$

These results show that production and dissipation of turbulent
kinetic energy decrease linearly with height. On the other
hand, we see that production and molecular destruction of
temperature variance is constant.

Next we turn to the profile of the σ_w, σ_u and σ_v which are
the standard deviations of the vertical, the longitudinal and
the horizontal velocity fluctuations. In addition we consider
the standard deviation of the temperature fluctuations, σ_θ. To
obtain explicit profiles for these quantities we use the
results of local scaling (Nieuwstadt, 1984), which gives
$\sigma_w \sim \|\tau\|^{\frac{1}{2}}$, $\sigma_u \sim \|\tau\|^{\frac{1}{2}}$, $\sigma_v \sim \|\tau\|^{\frac{1}{2}}$ and $\sigma_\theta \sim -\overline{w\theta} \,/\, \|\tau\|^{\frac{1}{2}}$. With the
aid of (12) and (14) we then find

$$\sigma_w/u_* = c_w \, (1 - z/h)^{3/4}, \qquad (26a)$$

$$\sigma_u/u_* = c_u \, (1 - z/h)^{3/4}, \qquad (26b)$$

$$\sigma_v/u_* = c_v \, (1 - z/h)^{3/4}, \qquad (26c)$$

$$\sigma_\theta/T_* = c_\theta \, (1 - z/h)^{1/4}, \qquad (26d)$$

These profiles are compared with experimental data in the Figs.
7 - 10. We see that observations for σ_w, σ_u and σ_v are
reasonably well described by (26a), (26b) and (26c) with
$c_w = 1.4$, $c_u = 2.04$ and $c_v = 1.7$. The results are less clear
for σ_θ. The observations show an increasing value of σ_θ in the
lower part of the boundary layer, whereas in the upper part of
the boundary layer the observed σ_θ shows the same tendency as
given by (26d). On the other hand a comparison between the
observation of the velocity and temperature variance clearly
indicates that the former decreases faster with height than the
latter. This result may reflect the behaviour of the production terms (24a) and (24b), according to which the production
of turbulent kinetic energy decreases with height whereas the
production of the temperature fluctuations remains constant.

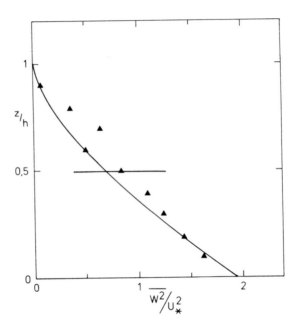

Fig. 7   The standard deviation of the vertical velocity fluctu-
         ations non-dimensionalized with u_* as a function of
         non-dimensional height.   For an explanation of the
         symbols see Fig. 1.   The solid line is the equation

$$\sigma_w/u_* = 1.4 \, (1 - z/h)^{3/4}.$$

As the last subject we consider the structure functions C_T^2
and C_V^2, which are important for acoustical and optical wave
propagation in the boundary layer.   Kaimal (1973) shows that
these structure functions can be related to ε and ε_θ by

$C_T^2 = 3.2 \, \varepsilon_\theta^{-1/3}$ and $C_V^2 = 2 \, \varepsilon^{2/3}$.   With the aid of (25a) and
(25b) we then find

$$\frac{C_T^2 \, (kh)^{2/3}}{T_*^2} = 3.2 \, \frac{Ri}{Ri_f^{5/3} (1-Ri_f)^{1/3}} \, (\frac{h}{L})^{2/3} \, (1 - \frac{z}{h})^{-1/3} \quad (27a)$$

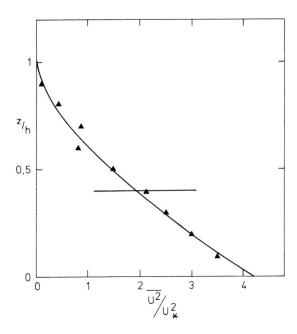

Fig. 8   The standard deviation of the longitudinal velocity
         fluctuations non-dimensionalized with u_* as a function
         of non-dimensional height. For an explanation of the
         symbols see Fig. 1.   The solid line is the equation
         $\sigma_u/u_* = 2.04 \ (1 - z/h)^{3/4}$.

$$\frac{C_V^2 \ (kh)^{2/3}}{u_*^2} = 2 \ \frac{(1 - Ri_f)^{2/3}}{Ri_f} \ (\frac{h}{L})^{2/3} \ (1 - \frac{z}{h})^{2/3} \qquad (27b)$$

We see that C_T^2 increases as a function of height.   The
signal of a monostatic acoustic sounder depends primarily on
C_T^2.   Therefore, this result may explain why such an instrument
is well suited to measure the height of the stable boundary
layer.

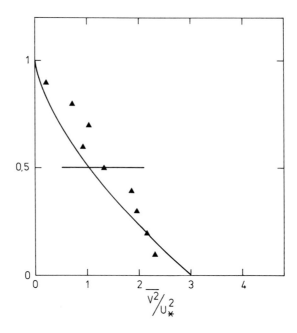

Fig. 9   The standard deviation of the lateral velocity fluctuations non-dimensionalized with u_* as a function of
non-dimensional height. For an explanation of the
symbols see Fig. 1. The solid line is the equation
$\sigma_v/u_* = 1.7 \ (1 - z/h)^{3/4}$.

(d)   The upper boundary

The topic of this subsection is the singularity that
appeared in our solution for $z \to h$. A good example is the
temperature profile (21). At the top of the boundary layer the
temperature must reach a value θ_h. This temperature is

independent of time because the turbulent heat flux is zero
above the boundary layer and because we have neglected other
processes which may change the temperature such as e.g. long
wave radiation. However, the logarithmic behaviour of the
temperature profile (21) leads to a meaningless value of θ_h.
Clearly, this is an undesirable feature.

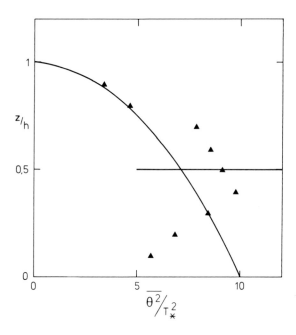

Fig. 10   The standard deviation of the temperature fluctuations
          non-dimensionalized with T_* as a function of non-
          dimensional height.   For an explanation of the symbols
          see Fig. 1.   The solid line is the equation

$$\sigma_\theta / T_* = 3.2 \, (1 - z/h)^{\frac{1}{4}}.$$

   To explain this singular behaviour we first note that the
model (5a)-(5d) cannot satisfy a stationary solution near
z = h.   Namely, we have found that stationarity implies a
linear temperature decrease with time.   Furthermore, we have
argued that the closure hypothesis (1a) and (1b) lead in case
of a stationary boundary layer to a temperature gradient which
is independent of time.   Consequently, the temperature profile
which satisfies these constraints, must have the form
$\Theta = f(z) + ct$, where c is a constant.   This is in agreement
with our temperature profile (21).   However, it is clear that
such a profile cannot satisfy a constant temperature at z = h.
Consequently, we should investigate the general solution of
(5a)-(5d) in order to describe the boundary layer structure
near its top.   Another complicating factor may be the fact that
our turbulence model (1a) and (1b) is no longer valid close to
z = h, because we have argued that the Richardson number

increases discontinuously across $z = h$. Such a discontinuity
must be included in the complete solution. In terms of our
turbulence model this result implies that the balance between
local production and destruction can no longer be satisfied.
Other factors such as the turbulent transport terms must
become important. A more detailed description of turbulence
near $z = h$ is not yet available. Therefore, a treatment of the
boundary layer near $z = h$ must await further developments.

However, let us assume as a first approximation that our
solution is valid in a region below a distance Δ from the
boundary-layer top. Furthermore, we hypothesize that $\Delta \simeq \ell_{max}$,
where ℓ_{max} is the maximum value of the turbulence length scale.
With the aid of (4) and (16b) we find $\ell_{max} = 0.5 \, k \, Ri_f / Ri^{\frac{1}{2}} \, L$
so that

$$\Delta = \alpha \, L \tag{28}$$

with $\alpha \simeq 0.1$. So according to our assumption the results,
derived above, apply to $z < h - \alpha \, L$. Let us look at some consequences. First we consider the temperature near the top of
the boundary layer. With $\Theta_h = \Theta(h - \alpha \, L)$ we find from (21)

$$\frac{\Theta_h - \Theta_0}{T_*} = \frac{Ri}{k \, Ri_f^2} \, \frac{h}{L} \, \ln(\alpha^{-1} \frac{h}{L}) . \tag{29}$$

This result is commonly denoted as a heat transfer law
(Zilitinkevich, 1975). It is equivalent to the resistance law
(18a). However, the dependence on the stability parameter h/L
is quite different from (18a).

To end this section we calculate the bulk Richardson number,
Ri_B, which is usually defined as

$$Ri_B = \frac{\frac{g}{T} (\Theta_h - \Theta_0) \, h}{\| w_g \|^2} . \tag{30}$$

Substitution of (18a) and (29) leads to

$$Ri_B = Ri \, \ln (\alpha^{-1} \frac{h}{L}) . \tag{31}$$

We find that Ri_B varies as a function of stability, despite the fact that the local Richardson number is a constant. This result is in agreement with the model calculations of Brost and Wyngaard (1978, 1979) which also exhibit an increase of Ri_B with h/L. It contrasts with other approaches, which take the bulk-Richardson number to be constant (e.g. Mahrt, 1981).

5. SUMMARY AND CONCLUSIONS

An analytic approach to the steady-state stable boundary layer has been developed. It is based on the hypothesis that the Richardson number and the flux-Richardson number are constant throughout the boundary layer.

The analysis leads to an equation for the boundary-layer height which is equal to the well-known expression proposed by Zilitinkevich. In addition we propose a theoretical value for the constant in this expression. The main result of our solution consists of explicit profiles for the temperature flux and the shear stress. From these we are able to calculate the mean temperature and the wind velocity as a function of height. These profiles reduce to the well-established Monin-Obukhov similarity forms for $z \to 0$. We also obtain expressions for the terms in the energy and temperature variance budgets. As a result we propose profiles for the standard deviations σ_w, σ_u, σ_v and σ_θ. We also consider the structure functions C_T^2 and C_V^2. A comparison with observations indicates that our solution agrees with many characteristics of the real stable boundary layer. At the top of the boundary layer we find a singularity. Its principal cause is a failure of the assumption of stationarity in this region. Furthermore, we suspect that turbulence can no longer be described here by a constant Richardson and flux Richardson number. We hypothesize a value for the height to which our solution remains valid. As a result we obtain an expression for the heat transfer law and the bulk Richardson number. The latter turns out to be not a constant but an increasing function of stability.

ACKNOWLEDGEMENT

This study was performed while the author was a visiting scientist at the University of Washington. The paper reflects comments of and discussions with R.A. Brost, J.A. Businger, J.W. Deardorff, J.C.R. Hunt, L. Mahrt, H. Tennekes and J.C. Wyngaard.

APPENDIX A

Correction for the solution near the surface.

Our assumption until now has been that the closure hypotheses (1a) and (1b) are valid throughout the boundary layer. As we have said before, this means that we use expression (4) for ℓ everywhere and that we neglect the asymptotic behaviour $\ell \sim z$ for $z \to 0$. It is the purpose of this appendix to investigate the condition for which such an omission is justified. Furthermore, we shall calculate an explicit correction term for our previous solution.

We have seen that the steady-state stable boundary layer can be alternatively formulated in terms of a gradient transfer hypothesis. Let us modify the resulting exchange coefficient (23a) to include the proper behaviour for $z \to 0$. We propose

$$K_M = \frac{ku_* \, z \, (1 - z/h)^2}{1 + Ri_f^{-1} \, z/L} \tag{A1}$$

For small values of z/h, $K_M \to k \, u_* \, z/(1 + Ri_f^{-1} \, z/L)$, which agrees with the well-known Monin-Obukhov similarity form for the stable surface layer (Businger et al., 1971; Brost and Wyngaard, 1978). For $z \gg L$ (A1) approaches the quadratic expression (23a) for K_M.

Next we express the equation for the steady-state stable boundary layer in terms of a gradient transfer hypothesis. From the momentum equation (6b) we obtain

$$K_M \frac{\partial^2 \tau}{\partial z^2} - if\tau = 0 \tag{A2}$$

In this equation we substitute (A1) and at the same time change to the dimensionless variables (7a)-(7c). The result is

$$\frac{(1 - \eta)^2}{1 + \dfrac{L}{h} \dfrac{Ri_f}{\eta}} \frac{\partial^2 \sigma}{\partial \eta^2} - i \, \sqrt{3} \, \sigma = 0 \, , \tag{A3}$$

where we have also used the Zilintinkevich equation for the boundary-layer height (10) and (15). To complete (A3) we must supply boundary conditions. They read $\sigma = 1$ at $\eta = 0$ and $\sigma = 0$ at $\eta = 1$.

Equation (A3) contains the stability parameter h/L. Let us restrict our attention to large values of this parameter, $h/L \gg 1$. In that case, we can expand σ in terms of the following asymptotic series in L/h.

$$\sigma = \sigma_0 + \frac{L}{h} \sigma_1 + \ldots \ldots \tag{A4}$$

After substituting this series in (A3) and equating powers of L/h we find

$$(1 - \eta)^2 \frac{\partial^2 \sigma_0}{\partial \eta^2} - i \sqrt{3} \, \sigma_0 = 0 \tag{A5}$$

$$(1 - \eta)^2 \frac{\partial^2 \sigma_1}{\partial \eta^2} - i \sqrt{3} \, \sigma_1 = i \, Ri_f \, \sqrt{3} \, \frac{\sigma_0}{\eta} \tag{A6}$$

The boundary conditions become

$$\sigma_0 = 1, \quad \sigma_1 = 0 \quad \text{for} \quad \eta = 0$$

$$\tag{A7}$$

$$\sigma_0 = 0, \quad \sigma_1 = 0 \quad \text{for} \quad \eta = 1$$

The equation (A5) for σ_0 leads to the previously found solution for the stress profile (14). So the results of section 4 can be interpreted as the first term in an asymptotic series for $h/L \gg 1$. (See also the discussion at the end of section 2.)

With (14) for σ_0 we can find the solution for the first-order correction σ_1. The result which satisfies the boundary conditions reads

$$\sigma_1 = -i \sqrt{3} \, Ri_f \, [(1 - \eta)^{\frac{1}{2}} \, (3 + i \sqrt{3}) \int_0^\eta (1 - \eta')^{-3 - i \sqrt{3}} \, x$$

(A8)

$$\{ \int_{\eta'}^1 \frac{1}{\eta''} (1 - \eta'')^{1 + i \sqrt{3}} \, d\eta'' \} \, d\eta']$$

From this equation we can obtain in principle the structure of the first order correction.

For example, let us calculate the correction term for the resistance law (18). For that purpose we need an equation from which we can obtain the velocity profile. We use (5b) for a stationary boundary layer. In dimensionless form it reads

$$\frac{W_g - W}{u_*} = \frac{i}{c^2} \frac{h}{L} \frac{\partial \sigma}{\partial \eta} \, , \tag{A9}$$

where we have used the expression for the boundary-layer height (10).

In section (4b) we have seen that the resistance law results by considering the limit of (A9) for $\eta \to 0$. From (14) and (A8) we find

$$\left(\frac{\partial \sigma_0}{\partial \eta} \right)_{\eta \to 0} = 1/3 \, (3 + i \sqrt{3}) ,$$

(A10)

$$\left(\frac{\partial \sigma_1}{\partial \eta} \right)_{\eta \to 0} = -i \sqrt{3} \, Ri_f \, [-\ell n \, \eta + (1 + i \sqrt{3}) \int_0^1 \ell n \, \eta \, (1-\eta)^{i \sqrt{3}} d\eta] ,$$

where we have applied intergration by parts to obtain the expression for $\partial \sigma_1 / \partial \eta$. Substitution in (A9) leads to

$$k \frac{W_g}{u_*} = \frac{1}{Ri_f} e^{-i \pi/3} \frac{h}{L} \dotplus \ell n(h/z_0) + (1 + i \sqrt{3}) \int_0^1 \ell n \, \eta \, (1-\eta)^{i \sqrt{3}} d\eta$$

(A11)

where we have used (15) and the well-known surface-layer result $W = k^{-1} u_* \ln (z/z_0)$ for $z \to 0$. The parameter z_0 is the roughness length. We see that the correction to our previous result (18) contains the familiar logarithmic term $\ln (h/z_0)$.

The definite integrals in (A11) can be evaluated numerically. In component form the resistance law then becomes

$$k \frac{U_g}{u_*} = \frac{1}{2Ri_f} \frac{h}{L} + \ln \left(\frac{h}{z_0}\right) - 1.4 \qquad \text{(A12)}$$

$$k \frac{V_g}{u_*} = \frac{\sqrt{3}}{2Ri_f} \frac{h}{L} - 0.8. \qquad \text{(A13)}$$

REFERENCES

André, J.C. and Mahrt, L. (1982) The Nocturnal Surface Inversion and Influence of Clear-Air Radiative Cooling. *J. Atmos. Sci.*, **39**, 864-878.

Brost, R.A. and Wyngaard, J.C. (1978) A Model Study of the Stably Stratified Planetary Boundary Layer. *J. Atmos. Sci.*, **35**, 1427-1440.

Brost, R.A. and Wyngaard, J.C. (1979) Reply. *J. Atmos. Sci.*, **36**, 1821-1822.

Businger, J.A. (1982) Equations and Concepts. Ch. 1. Atmospheric Turbulence and Air Pollution Modelling. F.T.M. Nieuwstadt and H. van Dop, (eds.), Reidel Publishing Company.

Businger, J.A., Wyngaard, J.C., Izumi, Y. and Bradley, E.F. (1971) Flux Profile Relationships in the Atmospheric Surface Layer. *J. Atmos. Sci.*, **28**, 101-109.

Caughey, S.J., Wyngaard, J.C. and Kaimal, J.C. (1979) Turbulence in the Evolving Stable Boundary Layer. *J. Atmos. Sci.*, **6**, 1041-1052.

Delage, Y. (1974) A Numerical Study of the Nocturnal Atmospheric Boundary Layer. *Quart. J. R. Met. Soc.*, **90**, 260-265.

Einaudi, F. and Finnigan, J.J. (1981) The Interaction between an Internal Gravity Wave and the Planetary Boundary Layer.

Part I: The linear analysis. *Quart. J.R. Met. Soc.*, **107**, 793-807.

Finnigan, J.J. and Einaudi, F. (1981) The Interaction between an Internal Gravity Wave and the Planetary Boundary Layer, Part II: Effect of the Wave on the Turbulence Structure. *Quart. J.R. Met. Soc.*, **107**, 807-832.

Garratt, J.R. (1981) Observations in the Nocturnal Boundary Layer. *Bound.-Layer Meteor.*, **22**, 21-48.

Garratt, J.R. and Brost, R.A. (1981) Radiative Cooling Effects within and above the Nocturnal Boundary Layer. *J. Atmos. Sci.*, **30**, 2730-2746.

Hunt, J.C.R. (1982) Diffusion in the Stable Boundary Layer. Ch. 6. Atmospheric Turbulence and Air Pollution Modelling. F.T.M. Nieuwstadt and H. van Dop (eds) Reidel Publishing Company.

Lettau, H.H. and Dabberdt, W.F. (1970) Variangular Wind Spirals. *Bound.-Layer Meteor.*, **1**, 64-79.

Mahrt, L. (1981) Modelling the Depth of the Stable Boundary-Layer. *Bound.-Layer Meteor.*, **21**, 3-19.

Mahrt, L., Heald, R.C., Lenshow, D.H. and Stankov, B.B. (1979) An Observational Study of the Structure of the Nocturnal Boundary Layer. *Bound.-Layer Meteor.*, **17**, 247-264.

Misra, P.K. (1979) Comments on "A Model Study of the Stably Stratified Planetary Boundary Layer". *J. Atmos. Sci.*, **36**, 1820-1821.

Nieuwstadt, F.T.M. and Tennekes, H. (1981) A Rate Equation for the Nocturnal Boundary-Layer Height. *J. Atmos. Sci.*, **38**, 1418-1428.

Nieuwstadt, F.T.M. (1984) The Turbulent Structure of the Stable Nocturnal Boundary Layer. To appear in *J.A.S.*, July issue.

Tennekes, H. (1982) Similarity relations, Scaling laws and Spectral Dynamics. Ch. 2 Atmospheric Turbulence and Air Pollution Modelling. F.T.M. Nieuwstadt and H. van Dop (eds), Reidel Publishing Company.

Tennekes, H. and Lumley, J.L. (1972) A First Course in Turbulence, the MIT Press.

Wyngaard, J.C. (1973) On Surface Layer Turbulence. Ch. 3 Workshop on Micrometeorology. D.A. Haugen (ed.), Amer. Meteor. Soc., Boston, Ma.

Wyngaard, J.C. (1975) Modeling the Planetary Boundary Layer Extension to the Stable Case. *Bound.-Layer Meteor.*, **9**, 441-460.

Zilitinkevich, S.S. (1972) On the Determination of the Height of the Ekman Boundary Layer. *Bound.-Layer Meteor.*, **3**, 141-145.

Zilitinkevich, S.S. (1975) Resistance Laws and Prediction Equations for the Depth of the Planetary Boundary Layer. *J. Atmos. Sci.*, **32**, 741-752.

PARAMETERIZATION OF THE LOW FREQUENCY PART OF SPECTRA OF HORIZONTAL VELOCITY COMPONENTS IN THE STABLE SURFACE BOUNDARY LAYER

S.E. Larsen

(Risø National Laboratory, Denmark)

and

H.R. Olesen

(Institute of Mathematical Statistics and Operations Research, Technical University of Denmark)

and

J. Højstrup

(Risø National Laboratory, Denmark)

ABSTRACT

Spectra of velocity components in the stable planetary surface boundary layer show two different regions: A high frequency 3D-turbulent part, which in the surface layer is well described by Monin-Obukhov similarity, and a low frequency part, which displays a strong increase with decreasing frequency. Here simple parameterizations for both regions are presented and discussed. The variation of the high frequency spectra of all three velocity components with stability can be described by the stability variation of one mixing length scale. The low frequency spectra of the two horizontal velocity components are found to be fairly well described by a $N^2 k^{-3}$ - law, where N is the Brunt-Väisälä frequency and k is the horizontal wave number. Finally the shortcomings of such a description is discussed and compared with another possible scheme used to described spectra for internal waves in the ocean.

1. INTRODUCTION

Spectra in the stable atmospheric surface layer are often described by use of parameters derived from the hypotheses of Monin-Obukhov, Kolmogorov and Taylor. Since we shall follow

this tradition, the relevant parameters and expressions are
summarized below.

   The velocity components are defined as u = $(\bar{u}+u', v', w')$,
where u, as indicated, is along the mean flow direction, v is
lateral and w vertical. Overbar denotes mean values, while '
describe quantities that average to zero. In the surface
layer the Monin-Obukhov hypothesis states that the important
parameters are the measuring height, z, the friction velocity,
$u* = (-\overline{u'w'})^{1/2}$, the heat flux, $\overline{w'\theta'}$, and the buoyancy parameter, g/T, where θ is the potential temperature, T the
absolute temperature and g the acceleration due to gravity.
Hence one can write the normalized velocity and temperature
gradients as

$$\frac{\kappa z}{u_*} \frac{\partial \bar{u}}{\partial z} = \phi_m(z/L) \sim 1 + 4.7 \ z/L$$

(1)

$$\frac{\kappa z \bar{u}_*}{\overline{w'\theta'}} \frac{\partial \bar{\theta}}{\partial z} = \phi_h(z/L) \sim 0.74 + 4.7 \ z/L,$$

where

$$L = -\frac{\kappa g}{T} \frac{u_*^3}{\overline{w'\theta'}}$$

is the Monin-Obukhov stability length and the expressions on
the right hand side pertain to stable conditions and are due
to Businger et al (1971). κ is the v. Karman constant (here
$\kappa \simeq 0.35$). Instead of z/L as a measure of stability, one can
use the Richardson number

$$Ri = \frac{g}{T} \frac{\partial \bar{\theta}/\partial z}{\left(\frac{\partial u}{\partial z}\right)^2} = \frac{z}{L} \phi_h/\phi_m^2$$ (2)

   For stationary flows at a sufficiently high Reynolds
number, such as is present in the atmosphere, the existence of
a local isotropy range in the large wave number part of the
velocity spectra is well established.

In the local isotropy range the spectra are functions of k, ε and ν only, where k is the wave number, ε the rate of dissipation of turbulent kinetic energy and ν the kinematic viscosity.

Let $\eta = (\nu^3/\varepsilon)^{1/4}$ be the Kolmogorov microscale. The part of the local isotropy range for which $k\eta \ll 1$ is called the inertial subrange, in which the velocity spectra, F(k), takes the familiar form

$$F_\gamma(k) = \alpha_\gamma \, \varepsilon^{2/3} \, k^{-5/3} \quad \text{with } \gamma = u,v,w \qquad (3)$$

where α_γ are the Kolmogorov constants.

Corresponding to (1) the Monin–Obukhov function for ε is given by:

$$\phi_\varepsilon = \phi_\varepsilon\left(\frac{z}{L}\right) = \kappa\varepsilon z/u_*^3 \sim \left\{1+5/2\,(z/L)^{2/3}\right\}^{3/2} \qquad (4)$$

where the right hand side is an empirical formulation due to Wyngaard and Coté (1972) and pertains to stable conditions only.

By use of (4) eq. (3) can be written

$$\frac{kF_\gamma(k)}{u_*^2 \, \phi_\varepsilon^{2/3}} = \alpha_\gamma \, (\kappa kz)^{-2/3}, \quad \gamma = u,v,w \qquad (5)$$

While most physical arguments about turbulence spectra are formulated in wave number space, most spectra are measured as frequency spectra, S(n). The transformation between the two variables is usually performed through use of Taylor's hypothesis of frozen turbulence, which relates frequencies n(Hz) to wave numbers and wave lengths, λ, as

$$\frac{k}{2\pi} = \frac{n}{u} = 1/\lambda = f/z \qquad (6)$$

where further the reduced frequency, f, has been defined. Hence (5) can be written

$$\frac{nS_\gamma(n)}{u_*^2 \, \phi_\varepsilon^{2/3}} = \frac{kF_\gamma(k)}{u_*^2 \, \phi_\varepsilon^{2/3}} = \frac{\alpha_\gamma}{(2\pi \, \kappa)^{2/3}} \; f^{-2/3} \tag{7}$$

which shows that the inertial subrange of velocity spectra in a Monin-Obukhov surface layer can be fully described by u_*, z/L and f.

For stable stratification the buoyancy forces extract energy from the turbulence for $k < k_b$, while the spectrum for $k\lambda_b \gg 1$ is largely unaffected. For $k\lambda_b < 1$ Lumley (1964) argues that the scalar energy spectrum, $E(k)$, can be written

$$E(k) \sim N^2 \, k^{-3} \tag{8}$$

where the Brunt-Väisälä frequency, N, is defined by

$$N = \left\{ \frac{g}{T} \frac{\partial \bar{\theta}}{\partial z} \right\}^{1/2} \tag{9}$$

and λ_b is given by

$$\lambda_b \sim (\varepsilon/N^3)^{1/2} \tag{10}$$

Weinstock (1978, 1980) emphasizes the importance of λ_b as well and derives a more complicated form of $E(k)$ than (8) with a gap in the spectrum and without a power law form of $E(k)$. Tchen (1975) considers a flow with velocity shear as well as temperature shear and derives a spectrum of the form

$$E(k) \sim f\left(\frac{\partial \bar{u}}{\partial z}, \; N\right) k^{-3} \tag{11}$$

The derivation of the expressions in (8, 11) are based on a number of assumptions which are fairly unrealistic for atmospheric surface layer turbulence. However, if a k^{-3}-law is observed they do offer suggestions of which parameter dependence to test. We have chosen to test the following forms.

$$F(k) = \begin{cases} \alpha \, N^2 \, k^{-3} \\[2em] \alpha \, \left(\dfrac{\partial u}{\partial z}\right)^2 k^{-3} \end{cases} \tag{12}$$

By use of the definitions of the variables, as given above, these forms can be written in surface layer scaled forms as

$$\frac{nS(n)}{u_*^2} = \begin{cases} \gamma \, \phi_h \, \dfrac{z}{L} \, f^{-2} = \dfrac{\alpha}{\kappa^2} \, Ri \, \phi_m^2 \, f^{-2} & \text{(a)} \\[2em] \gamma \, \phi_m^2 \, f^{-2} & \text{(b)} \end{cases} \tag{13}$$

where γ is a coefficient to be determined. From (13) is seen that the two different expressions in (12) will vary quite differently with stability.

Our extensive use of Taylor's hypothesis of frozen turbulence for low frequency spectra implies that the relation between theoretical predictions in wave number space and measurements interpreted in f-space may be somewhat hazy.

The conditions for Taylor's hypothesis to hold for stable to neutral shear flow have been formulated as (Powell and Elderkin, 1974, and Monin and Yaglom, 1975):

$$\sigma_u/\bar{u} \ll 1 \tag{14}$$

$$k\bar{u}/\frac{\partial \bar{u}}{\partial z} \gtrsim 4$$

The first of these will usually be fulfilled in the stable surface layer. We write the second as

$$f \gtrsim \phi_m/(\kappa u/u_*) = \left[1+4.7\frac{z}{L}\right]/\left[\ln\frac{z}{z_0} + 4.7\frac{z}{L}\right], \tag{15}$$

where we have used (1.6) and introduced the roughness length, z_0. Eq. (15) shows that if one uses Taylor's hypothesis for $f < 0.1$, one must expect to be on uncertain grounds.

2. 3D TURBULENCE SPECTRA

A simple spectral form in the stable surface layer has been suggested by Kaimal (1973).

$$\frac{nS_\gamma(n)}{\sigma_\gamma^2} = \frac{A_\gamma(f/f_0)}{1+A_\gamma \cdot (f/f_0)^{5/3}} \ , \ \gamma = u,v,w \tag{16}$$

$$\text{with } f_{0\gamma} = B_\gamma Ri + C_\gamma$$

Eq. (16) means that the spectral form is a bell shape that obeys Kolmogorov's inertial form for high frequencies and which moves to higher frequencies for increasing Ri. It further implies that the stability variation of the spectra of all three components can be considered controlled by the variation of a single length scale, see also Hunt (1982) and Hunt et al. (1983).

In Fig. 1 (taken from Olesen et al., 1983) we have presented spectral surface layer data in a slightly different form. The data were obtained during the Kansas 1968 experiment for a measuring height of 5.6 m and the analysis is described in Larsen and Busch (1976).

The data are organized to follow the form of (7) for high frequencies. Furthermore the peak is assumed to move in accordance with a simple mixing length description.

For the momentum flux in the surface layer we can derive an eddy diffusivity for momentum from

$$u_\star^2 = K_m \frac{\partial \bar{u}}{\partial z} \tag{17}$$

By use of (1) we obtain

$$K_m = u_\star \kappa z/\phi_m \sim \ell \, u_\star \tag{18}$$

where we on the right hand side have introduced the mixing length, ℓ, and as characteristic velocity have used u_*.   We now assume that the wavelength for which the spectra have their maximum is given by

$$\lambda_{m\gamma} = C_\gamma \ell, \quad \gamma = u,v,w, \tag{19}$$

where C_γ is a constant for each component.

From (19) we obtain, by use of (6)

$$f_m = f_{mN} \phi_m \tag{20}$$

where f_{mN} is the peak frequency for neutral conditions.

With these constraints Olesen et al. (1984) find the following expressions

$$\frac{nS(n)}{u_*^2} \left(\frac{q}{\phi_\varepsilon}\right)^{2/3} = \begin{cases} \dfrac{79x}{1+263x^{5/3}} & \text{for } u \\[3mm] \dfrac{13x}{1+32x^{5/3}} & \text{for } v \\[3mm] \dfrac{3.5x}{1+8.6x^{5/3}} & \text{for } w \end{cases} \tag{21}$$

with $x = f/q$ and $q = \phi_m(z/L)$

Equations (21) were chosen out of a more general family of curves given by

$$\frac{nS(n)}{u_*^2} \left(\frac{q}{\phi_\varepsilon}\right)^{2/3} = \frac{ax^\delta}{(1+bx^\alpha)^\beta} \tag{22}$$

with $\delta - \alpha\beta = -2/3$.

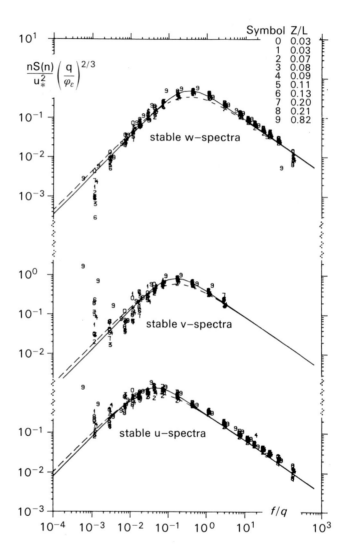

Fig. 1   Surface layer velocity spectra measured 5.6 m above terrain for different values of z/L, organized as described in the text with $q = \phi_m$ and with the unbroken curves describing eq. (21). The broken curves describe the corresponding fit to eq. (22).

The broken curves on Fig. 1 show the fit to the data for $\delta = \alpha = 1$. This form clearly fits the data less than (21).

The expressions in (21) are seen to give quite a good description of the data, which hereby are seen to conform quite well to a whole set of hypotheses: The Monin-Obukhov scaling laws, the Kolmogorov local inertial hypothesis, Taylor's frozen turbulence hypothesis and the mixing length assumption.

The data fit on Fig. 1 is performed for data measured at 5.6 m only. However, it is worth noting, that the data presented in Caughey et al. (1979), showing λ_m/h and $\varepsilon h/u_*^3$ vs. z/h, indicate that the expressions in (21) appear to be valid for $z/h \lesssim 0.1$. Here h is the height of the stable boundary layer. It must be emphasized, however, that the data of Caughey et al. are filtered, while ours are not. Hence, the above results based on a comparison between the two data sets, must be considered tentative.

A closer inspection of Fig. 1 shows some deviations between the data and (21). We notice: a) Especially for the w-spectra the high frequency part of the most stable spectra decreases faster than the -5/3-law. b) At low frequencies the data indicate a somewhat steeper ascent towards the peak for increasing frequency, than indicated by the +1-slope of (21). c) The lowest frequencies of the u- and v-spectra show a spectral region with a steep slope, that obviously does not follow either the form or the scaling laws used in (21).

These phenomena shall be discussed below.

The width of the inertial range.

In section 1 the Kolmogorov scale, was mentioned. This scale can be used to estimate the largest wave numbers still within the inertial range of the spectrum. For one dimensional spectra $k.\eta \sim 0.1$ marks the transition from the inertial range to the faster decrease of the dissipation subrange. This means that the upper frequency for the inertial subrange, f_{ui}, can be found as

$$f_{ui} = z/(20\pi\eta) = \frac{\kappa^{-1/4}}{20\pi} \phi_\varepsilon^{1/4} \left(\frac{z\,u_*}{\nu}\right)^{3/4} \tag{23}$$

Correspondingly the lower frequency for the inertial subrange,
f_{li}, can be found as (Busch and Larsen, 1972 and Kaimal et al.,
1972)

$$f_{li} \sim 10 \ f_m = 10 \ f_{mN} \ \phi_m \tag{24}$$

From (23) and (24) the width of the inertial range can be
described by f_{ui}/f_{li}. For the w-spectra, we find for the most
neutral data on Fig. 1 a ratio of 40 while the most stable
data give a ratio around 4. As pointed out by Busch and Larsen
(1972) the inertial subrange might therefore be virtually
absent for very stable surface layer spectra.

Low frequency part of turbulence spectra.

On the low frequency side of the peak the spectra of all 3
velocity components show a tendency to decrease faster with
decreasing frequency than indicated by the model. This
tendency is enhanced with growing z/L.

The attenuation of the low frequency part of the spectrum
might be seen as a continuation of the change in spectral
shape, that happens when z/L goes from weakly unstable to
weakly stable.

This latter change is illustrated on Fig. 2, which is taken
from the same data as Fig. 1, but includes near neutral
unstable data as well.

Fig. 2 shows the so called "forbidden" region between the
stable and the unstable spectra of horizontal velocity
components in the surface layer as originally noticed by Kaimal
et al. (1972) and Busch and Larsen (1972).

The low frequency behaviour of the spectra for unstable
situations has recently been modelled by Kaimal (1980) and
Højstrup (1982). In both papers the low frequency behaviour
of the spectra is attributed to eddies whose size is of the
order of the boundary layer depth and whose intensity scales
with the parameters describing the turbulence in the total
boundary layer (i.e. for unstable situations mixed layer
scaling). Fig. 2 shows how the model of Højstrup accurately
predicts the spectrum for z/L as close to neutral as -0.04.

However, also for neutral and (at least weakly) stable
conditions the planetary boundary layer is known to contain
eddies, that scale with the boundary layer height (Tennekes,

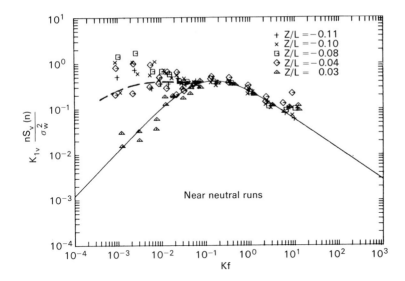

Fig. 2   Near neutral v-spectra (Larsen and Busch, 1976) the
         data are collapsed on the neutral spectral shape by a
         method as the one used in Fig. 1 (Busch, 1973).
         The broken curve corresponds to the model prediction
         by Højstrup (1982) with z/L = -0.04 and z_i = 1000 m.

1973, Brown, 1972. See also the paper by King in this
volume). Therefore one might expect that spectra for neutral
to stable conditions behave similarly to unstable spectra.
However, in stable conditions turbulence is damped by buoyancy
and also the typical boundary layer height is smaller than
for unstable stratification.

The latter phenomenon means that eddies, which scale with
the boundary layer height, will show up at higher f-values and
thereby closer to the peak of the surface layer scaled spectrum
for stable than for unstable stratifications. If the boundary
layer eddies show up close enough to the peak, this might
result in an increased steepness of the low frequency behaviour
of the spectrum. Furthermore, this argument indicates that the
neat surface layer forms in Fig. 1 might hide a contribution,
which scales with the boundary layer height, h, a contribution
which will be very difficult to sort out, because h in measurements correlate with L (see Caughey et al. (1979), Caughey
(1982)).

To estimate the importance of the buoyancy damping we might use λ_b from (10), which can be considered the smallest wavelength for buoyancy dominance of the spectrum. By use of the similarity forms for ε and N we can write:

$$\lambda_b = \kappa z \, \phi_\varepsilon^{1/2}/(\phi_h \, z/L)^{3/4} \sim \begin{cases} \kappa z^{1/4} \, L^{3/4} & , \; z/L \ll 1 \\[2ex] \kappa L & , \; z/L \gg 1 \end{cases} \tag{25}$$

From (20) we have $f_m = f_{mN} (1+4.7 \, z/L)$. Therefore equation (25) shows that $f_b = z/\lambda_b$ increases faster with z/L than f_m, until z/L becomes large enough for both f_m and f_b to increase proportionally to z/L. For z/L increasing from zero, we therefore expect the low frequency slope of the spectrum to increase, until it reaches a limiting value when both f_m and f_b increase proportionally to z/L. In Fig. 3 this is illustrated in terms of the variation of estimates of the parameter δ, from eq. (22), for the data on Fig. 1. It is seen that δ reaches a limiting value between 1.1 and 1.2 for z/L values around 0.1-0.2, which are roughly the values where both f_m and f_b start varying proportionally to z/L. It should be noted that the δ-estimates are fairly uncertain. For this reason and because δ is fairly close to 1 anyhow, we have not entered this additional complication into the expressions used to fit the data on Fig. 1.

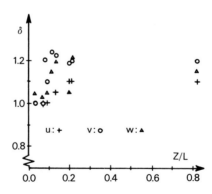

Fig. 3   Estimates of the low frequency slope, δ (eq. 22), as function of z/L for the three component spectra of Fig. 1.

Equation (25) can be used to estimate when the transition from an unstable-neutral to a stable-neutral spectral shape occurs for increasing stability. We assume that it occurs when λ_b becomes smaller than eddies of the size of the boundary layer height. This can be tested for two standard expressions for the near neutral to stable boundary layer height (f_c is the Coriolis parameter).

$$h \sim \begin{cases} 0.3 \ u_*/f_c \\ 0.7 \ (Lu_*/f_c)^{1/2} \end{cases} \tag{26}$$

If we by virtue of (25) let $\lambda_b \sim \kappa L$, it is seen that the boundary layer size eddies are damped for $\mu \simeq 1$, μ being the Monin-Kazansky stability parameter ($\mu = \kappa u_*/(Lf_c)$, Zilitinkevich, 1975). This corresponds to stable conditions but very close to neutral.

At the end of this discussion, it could be added that the low frequency behaviour of a "true" neutral spectrum was considered fairly academic, when the neutral shape of the horizontal velocity components were originally established. However, development of building and construction techniques, where the structures show resonance response in the frequency region 0.001-1 Hz, has made an assessment of the "true" spectrum important in this region to establish design standards, as is obvious from Fig. 2, where the intensity will vary one to two orders of magnitude depending on whether one assumes a "true" neutral spectrum to be close to an unstable-neutral or a stable-neutral spectrum.

The above arguments indicate that a "true" neutral spectrum is probably closest to an unstable-neutral form.

3. LOW FREQUENCY HORIZONTAL SPECTRA

In section 2 it was noted that the spectral expressions in (21) did not describe the behaviour at the lowest frequencies of the u- and v-spectra. As described in section 1 models exist to describe the form of this frequency range if the spectra are found to follow a f^{-3}-law. Therefore this was tested and the estimated power law was found to be a little less than 3 but not smaller than 2.5 (Olesen et al., 1984), but given that our frequencies were close to the bottom of the gap between the low frequency region and the 3D turbulence region described in section 2, this was considered not to be

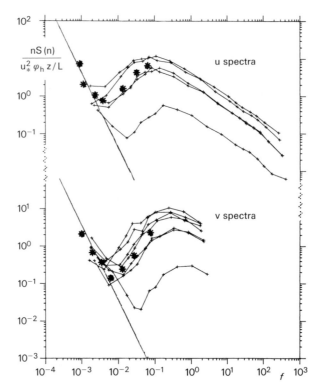

Fig. 4   U- and v-spectra from Fig. 1 scaled in accordance
         with (27) (Olesen et al., 1984).   The dots represent
         the stable North Sea spectra from Fig. 5.   The most
         neutral spectra, showing no tendency of a f^{-2}-law,
         are not shown on the plot.

inconsistent with f^{-3}-behaviour.   In Fig. 4 the low
frequency part of the spectra of the horizontal components
have been plotted, when scaled in accordance with the behaviour
predicted by (8).

   The lines shown on the plot correspond to

$$\frac{nS_\gamma(n)}{u_*^2} = \gamma \, z/L \, \phi_h(z/L) \, f^{-2}, \quad \gamma = u,v \qquad (27)$$

with $\gamma = 4.10^{-6}$.

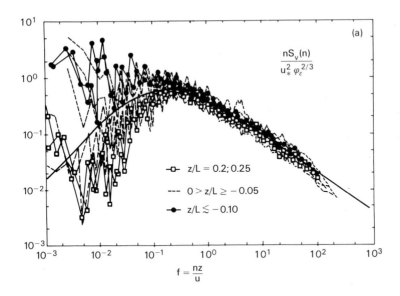

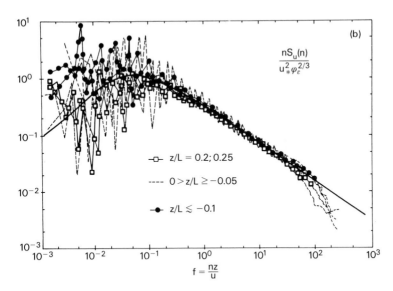

Fig. 5 Spectral data obtained 7 metres above the water in the
JONSWAP-1975 North-Sea-Experiment (Rasmussen et al.,
1981). The data are organized in accordance with (7).
The curves describe the neutral surface layer forms
according to Kaimal et al. (1972).

This is seen to give a fairly good description of the data.
We have not shown a fit comparing (13b) to the data. However,
(13) shows that if (13a) works for the data then (13b) can
not work, since Ri varies from 0.02 to 0.2 in the data set.

Fig. 4 illustrates that the u-spectra show less of an f^{-2}-
law in the frequency range considered. The reason for this is
that the 3D turbulence part of the u-spectrum is displaced
towards lower frequencies than for the v-spectra. From the
figure there is no justification for assuming a different
coefficient in (27) for u- and v-spectra. On the plot we have
furthermore included a stable spectrum measured over the North
Sea in the JONSWAP-1975 experiment. The total set of u- and
v-spectra from this experiment is shown in Fig. 5.

Additional data.

Obviously, the data set used in Fig. 4 is somewhat limited.
In Fig. 6 two additional stable spectral data set are
presented. The spectra in Fig. 6a were measured in night time
conditions in central Jutland, Denmark, while the u-spectrum
in Fig. 6b was measured at night at a coastal station on the
Swedish Baltic island Gotland. The latter data were obtained
in May when the surrounding water was cold. Relevant parameters for the two data sets are presented in Table 1.

Table 1

The table presents relevant parameter values for the spectra in
Figs. 6a and 6b.

Figure	u^* (m/s)	$\bar{u}$ (m/s)	z (m)	z/L	Run length (hr)
6a	0.055	3	10	1	1
6b	0.5	6	10	0.02	7

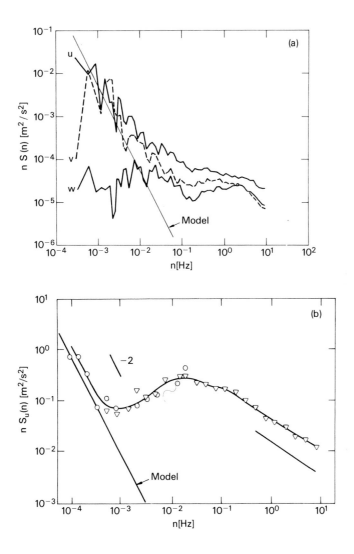

Fig. 6 Stable surface layer data obtained from a: T. Mikkelsen
(personal communication) and b: A.-S. Smedman and
U. Högström (personal communication). The lines
denoted "Model" are computed from (27) and the relevant
parameter values from Table 1. Fig. 6b is composite:
Cup-anemometer readings over 8 hours: O; turbulence
data over the central one hour: Δ.

From the parameter values the model predictions of (27) have
been computed. From Fig. 6 it is seen that the model fits
the data well.

Position of the gap

   Högström and Högström (1975) find that the gap frequency,
f_{gap}, varies with z/L. Our data as well as data from Zhou and
Panofsky (1983) show such a relationship, essentially
indicating a linear relation between f_{gap} and z/L. The low
frequency behaviour of the spectra in (21) can be written:

$$nS_\alpha(n)/u_*^2 = A_\alpha \; \phi_\varepsilon^{2/3} \; \phi_m^{-5/3} \; f \qquad (28)$$

with A_v = 13 and A_u = 79.

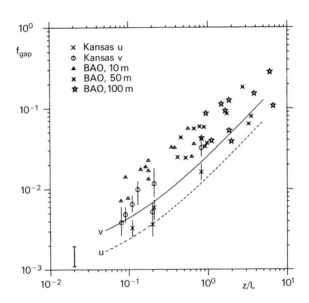

Fig. 7   The variation of the gap frequency, f_{gap}, with z/L.
The data denoted BAO (Boulder Atmospheric Observatory)
are taken from Zhou and Panofsky (1983) and represent
v-spectra. For the Kansas data the beams indicate
uncertainty range. The heavy beam shows the position
of the gap in Fig. 6b.

If we assume that the spectra in the gap can be written as a sum of (28) and (27) we obtain

$$f_{gap,\alpha} = [2(\gamma/A_\alpha) \; (z/L) \; (\phi_h \cdot \phi_m) \; (\phi_m/\phi_\varepsilon)^{2/3}]^{1/3} \qquad (29)$$

which varies as $(z/L)^{1/3}$ for small z/L and as z/L for large z/L.

From Fig. 7 is seen that a simple proportionality between f_{gap} and z/L gives a better description of the data, than does (29). This is not really surprising since (28), as shown in Fig. 3, gives a fair description of the low frequency turbulence data only. Fig. 7 indicates a possibility of systematic difference between the Kansas data and the data from BAO. At present it is unknown if this difference is due to differences in analysis technique or in terrain forms or a stratification with z for small z-values.

Other aspects of the low frequency spectra

Many studies have found that the low frequency spectra measured throughout the stable boundary layer change very little with height when plotted as function of n (Hz), (Caughey, 1977, 1982 and Zhou and Panofsky, 1983). This phenomenon is illustrated in Fig. 8, taken from Zhou and Panofsky (1983).

The model description in (27) can be written as:

$$\frac{ns(n)}{u_*^2} \sim \begin{cases} \dfrac{\bar{u}^2}{Lz} \; n^{-2} & z/L \ll 1 \\[4mm] \dfrac{\bar{u}^2}{L^2} \; n^{-2} & z/L \gg 1 \end{cases} \qquad (30)$$

which is roughly independent of z for $z/L < 1$ while a fairly strong variation with z as $\bar{u}^{-2}$, is indicated for $z/L > 1$. However, it should be pointed out that the results of Finnigan and Einaudi (1981) and Hunt et al. (1983) indicate that some variation of the spectrum with z must be expected.

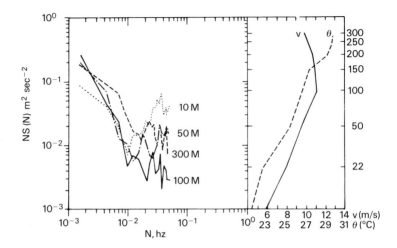

Fig. 8 The constancy of the low frequency v-spectra with
        height within the boundary layer, shown together with
        the mean velocity and temperature profiles (BAO-data
        from Zhou and Panofsky (1983)).

Finally we mention that the low frequency spectrum nS(n) is
found often to have its peak around 10^{-3} Hz (Caughey, 1977,
1982), but other data fail to find such maxima, e.g. Fig. 6b,
and the problems associated with stationarity for such low
frequencies make estimates of nS(n) for n < 10^{-3} fairly
tentative.

DISCUSSION

    Power spectra of the horizontal velocity components in the
stable atmospheric surface layer have been discussed.

    The spectra were found to separate into a high frequency
region, essentially being neatly described by the Monin-
Obukhov formalism (21), and into a low frequency region, that
was found to depend on the Brunt-Väisälä frequency, as
described by (27).

    Although (27) was shown to be a good predictor of the low
frequency spectra for the data sets studied (surface layer
spectra over fairly flat terrain at mid-latitude), it has a
number of deficiencies from a theoretical point of view. The
fact that (27) does not contain a parameter describing the

energy involved in the spectrum suggests that some kind of
unknown saturation process must be involved. The "universal"
coefficient in the equation is very far from being of order
unity, which further indicates that essential physical parameters have not been included in the formula. Finally it is
not clear to which degree the spectrum is the result of turbulence and/or gravity waves. Lumley (1964) argues that this is
more a problem of semantics than of substance. However, the
fact that we have employed Taylor's hypothesis far outside its
expected frequency range of validity, as well as the reports on
the obvious importance of gravity waves in the stable boundary
layer (Einaudi and Finnigan, 1981, and Finnigan and Einaudi,
1981) indicate that the role of gravity waves have to be
included in a more detailed description. Therefore it is
interesting to compare (27) with the results of Garret and
Munk (1972, 1975) and Munk (1981). They describe the spectra
of internal gravity waves in the ocean assuming an exponential
variation of the Brunt-Väisälä frequency

$$N = N_O \, e^{-z/b} \tag{31}$$

Their spectral modelling is mostly concerned about evaluation
of the interplay between the dispersion relation for gravity
waves and the varying Brunt-Väisälä frequency and the resulting
demands to the spectral forms. Their result can be presented
as (Munk, 1981)

$$F_u(j(k,\omega),\omega) = \frac{2}{\pi} E \, f_c b^2 \, N_O \, N \, \omega^{-3} \, (\omega^2 + f_c^2)(\omega^2 - f_c^2)^{-1/2} H(j) \tag{32}$$

where $H(j)$ is a function of mode number, j, that integrates
to 1, leading to

$$F_u(\omega) = \frac{2}{\pi} E \, f_c b^2 \, N_O \, N \, \omega^{-2} \quad \text{for } \omega \gg f_c. \tag{33}$$

Since the exact power law in (32, 33) is empirical, eq. (33)
appears fairly close to what can be deduced from (27). The
main difference is that (33) includes a parameterization of
the whole layer of interest, while (27) is strictly local.
In this respect (33) appears most satisfying in our opinion.

Equation (33) shares the disadvantage with (27) that it must appeal to an unknown saturation process to describe the spectral intensity. In this connection it is interesting to notice that the universal number, E, describing this intensity in (32) comes out as $\sim 6.3.10^{-5}$ being almost as small as in (27). As it is now, eq. (32) is not well suited to describe the stable atmospheric boundary layer, because the height dependence of N in the stable boundary layer is badly described by (31) and because the role of velocity shear is not included. However, the above arguments point to that the approach used to describe internal waves in the ocean could be found fruitful for a more detailed description of the stable low frequency atmospheric boundary layer spectra as well.

REFERENCES

Brown, R.A. (1972) On the inflection point instability of a stratified Ekman boundary layer. *J. Atmos. Sci.*, **29**, 850-859.

Busch, N.E. and Larsen, S.E. (1972) Spectra of turbulence in the atmospheric surface layer. In: Risø Report No. 256, 187-207.

Busch, N.E. (1973) On the mechanics of atmospheric turbulence. In: Workshop in Micrometeorology. (D.A. Haugen (ed.)). Amer. Meteorol. Soc. Boston, USA, 1-65.

Businger, J.A., Wyngaard, J.C., Izumi, Y. and Bradley, E.F. (1971) Flux-profile relationships in the atmospheric surface layer. *J. Atmos. Sci.*, **28**, 181-189.

Caughey, S.J. (1977) Boundary-layer turbulence spectra in stable conditions. *Boundary-Layer Meteorol.*, **11**, 3-14.

Caughey, S.J., Wyngaard, J.C. and Kaimal, J.C. (1979) Turbulence in the evolving stable boundary layer. *J. Atmos. Sci.*, **36**, 1040-1052.

Caughey, S.J. (1982) Observed characteristics of the atmospheric boundary layer. In: Atmospheric Turbulence and Air Pollution Modelling. (F.T.M. Nieuwstadt and H. van Dop (eds.)), D. Reidel, Dordrecht, Holland, 107-158.

Einaudi, F. and Finnegan, J.J. (1981) The interaction between an internal gravity wave and the planetary boundary layer. Part I. The linear analysis. *Quart. J. R. Met. Soc.*, **107**, 793-806.

Finnegan, J.J. and Einaudi, F. (1981) The interaction between
    an internal gravity wave and the planetary boundary layer.
    Part II: Effect of the wave on the turbulence structure.
    Quart. J. R. Met. Soc., **107**, 807-832.

Garret, C. and Munk, W. (1972) Space-time scales of internal
    waves. *Geophys. Fluid Dynamics,* **2**, 225-264.

Garret, C. and Munk, W. (1975) Space-time scales of internal
    waves. *J. Geophys. Res.,* **80**, 291-297.

Hunt, J.C.R. (1982) Diffusion in the stable boundary layer.
    In: Atmospheric Turbulence and Air pollution modelling.
    (F.T.M. Nieuwstadt and H. van Dop (eds.)), D. Reidel,
    Dordrecht, Holland, 231-274.

Hunt, J.C.R., Kaimal, J.C., Gaynor, J.E. and Korrell, A. (1983)
    Observations of turbulence structure in stable layers of
    the Boulder Atmospheric Observatory. In: Studies of
    nocturnal stable layers at BAO (J.C. Kaimal, (ed.)),
    Report no. 4 NOAA/ERL, Boulder, Colorado 80303.

Högström, A.-S. and Högström, U. (1975) Spectral gap in
    surface layer measurements. *J. Atmos. Sci.,* **32**, 340-350.

Højstrup, J. (1982) Velocity spectra in the unstable planetary
    boundary layer. *J. Atmos. Sci.,* **39**, 2239-2248.

Kaimal, J.C., Wyngaard, J.C., Izumi, Y. and Cote, O.R. (1972)
    Spectral characteristics of surface-layer turbulence.
    Quart. J. R. Met. Soc., **98**, 563-589.

Kaimal, J.C. (1973) Turbulence spectra, length scales and
    structure parameters in the stable surface layer. *Boundary-
    Layer Meteorol.,* **4**, 289-309.

Kaimal, J.C. (1978) Horizontal velocity spectra in an unstable
    surface layer. *J. Atmos. Sci.,* **35**, 18-24.

Larsen, S.E. and Busch, N.E. (1976) Hot-wire measurements in
    the atmosphere - Part II - A field experiment in the surface
    boundary layer. DISA-Inf. No. 20, 5-21.

Lumley, J.L. (1964) The spectrum of nearly inertial turbulence
    in a stably stratified fluid. *J. Atmos. Sci.,* **21**, 99-102.

Monin, A.S. and Yaglom, A.M. (1975) Statistical fluid
    mechanics: Mechanics of turbulence Vol. 2. MIT Press, London,
    England, 873.

Munk, W. (1981)   Internal waves and small-scale processes.
   In: Evolution of Physical Oceanography (B.A. Warren and
   C. Wunsch, (eds.)) MIT Press, London, England, 264-291.

Olesen, H.R., Larsen, S.E. and Højstrup, J. (1984)   Modelling
   velocity spectra in the lower part of the planetary
   boundary layer. Boundary-Layer Meteorol., 29, 285-312.

Powell, D.C. and Elderkin, C.E. (1974)   An investigation of
   the application of Taylor's hypothesis to atmospheric
   boundary layer turbulence. J. Atmos. Sci., 31, 990-1002.

Rasmussen, K.R., Larsen, S.E. and Jørgensen, F.E. (1981)
   Study of flow deformation around wind-vane mounted three-
   dimensional hot-wire probes. DISA-Inf. No. 26, 27-34.

Smedman, A.-S. and Högström, U. (1983)   Turbulent characteristics of a shallow convective internal boundary layer.
   Boundary-Layer Meteorol., 25, 271-287.

Turner, J.S. (1973)   Buoyancy effects in fluids. Cambridge
   University Press, Cambridge, England.

Tchen, C.M. (1975)   Cascade theory of turbulence in a stratified medium. Tellus, 27, 1-14.

Tennekes, H. (1973)   Similarity laws and scale relations in
   planetary boundary layers. In Workshop in Micrometeorology (D.A. Haugen (ed.)), Amer. Meteorol. Soc.,
   Boston, USA, 177-216.

Weinstock, J. (1978)   On the theory of turbulence in the
   Buoyancy subrange of stably stratified flows. J. Atmos.
   Sci., 35, 634-649.

Weinstock, J. (1980)   A theory of gaps in the turbulence
   spectra of stably stratified shear flows. J. Atmos. Sci.,
   37, 1547-1549.

Wyngaard, J.C. and Coté, O.R. (1971)   The budgets of turbulent
   Kinetic energy and temperature variance in the atmospheric
   surface layer. J. Atmos. Sci., 28, 190-201.

Zhou, L. and Panofsky, H.A. (1983)   Wind fluctuations in
   stable air at the Boulder tower. Boundary-Layer Meteorol.,
   25, 353-362.

Zilitinkevich, S.S. (1975)   Resistance laws and prediction
   equations for the depth of the planetary boundary layer.
   J. Atmos. Sci., 32, 741-752.

MODELLING THE DEVELOPMENT OF LARGE EDDIES IN THE STABLE ATMOSPHERIC BOUNDARY LAYER

J.C. King

(Meteorological Research Unit, RAF Cardington, Bedford)

ABSTRACT

A two-dimensional numerical model in which small scale turbulence is parametrized by a mixing-length hypothesis has been used to study the development of eddies in a flow which resembles that observed in a stably-stratified (e.g. nocturnal) atmospheric boundary layer. At least two types of instability are observed to develop; firstly a short wavelength mode associated with the weak jet profile at the top of the boundary layer and secondly a longer wavelength trapped mode. The structure of the trapped mode is similar to that of the "resonant overreflection" modes found by Davis and Peltier (1976) in a simple model of stratified shear flow. The short wavelength mode is probably a Kelvin-Helmholtz instability and it would appear that the presence of this mode greatly increases the initial growth rate of the trapped mode. Neither mode, however, makes a significant contribution to the net transport of heat or momentum. The form of these modes of instability does not appear to be critically dependent on the values of constants used in the turbulence parametrization, but is greatly affected by the choice of vertical grid spacing used within the boundary layer.

INTRODUCTION

Observations of the stable boundary layer (e.g. Caughey and Readings (1975), Merrill (1977), Hunt et al. (1983)) have revealed a great variety of wave-like phenomena on various length scales. While some wave motions are undoubtedly generated by topographic variations or mesoscale disturbances, others appear to be due to a variety of shear flow instability mechanisms, some of which are only poorly understood.

Furthermore, the role played by waves in transport processes,
through interactions with small scale turbulence, has yet to be
quantified.

Theoretical studies of stratified shear flow instability
have, of necessity, been limited to fairly idealised profiles
of wind and buoyancy.   In particular, most workers have used a
constant viscosity throughout the flow whereas in the turbulent
atmospheric boundary layer we expect the (eddy) viscosity to be
a function of static stability, wind shear and distance from
the ground. The present study uses a model in which the
viscosity is determined from a simple turbulence parametrization appropriate to the stable boundary layer. The model
uses a two-dimensional computational domain and can thus only
resolve two-dimensional disturbances. All three-dimensional
motions and sub-grid scale turbulence must be accounted for by
the parametrization. There is no a priori justification for
assuming that two-dimensional modes will be the most unstable
or most energetic. Hence the limited scope of this study
should be recognised - it is not intended to be a complete
simulation of the stable boundary layer but rather an investigation into the possible importance of a certain class of
modes.

The model is based on that used by Mason and Sykes (1982)
to study horizontal roll vortices in the convective atmospheric
boundary layer. Some changes have been made to the turbulence
parametrization, this is described in detail below, but the
numerical scheme is largely unaltered and the reader is
referred to Mason and Sykes (1982) for full details.

DESCRIPTION OF THE MODEL

The model solves the two-dimensional equations of motion for
an incompressible, rotating Boussinesq fluid:

$$\frac{\partial u}{\partial t} + u\frac{\partial u}{\partial x} + w\frac{\partial u}{\partial z} = -\frac{\partial p}{\partial x} - \frac{\partial P_O}{\partial x} + fv + \frac{\partial \tau_{11}}{\partial x} + \frac{\partial \tau_{13}}{\partial z} \tag{1}$$

$$\frac{\partial v}{\partial t} + u\frac{\partial v}{\partial x} + w\frac{\partial v}{\partial z} = -\frac{\partial P_O}{\partial y} - fu + \frac{\partial \tau_{12}}{\partial x} + \frac{\partial \tau_{23}}{\partial z} \tag{2}$$

$$\frac{\partial w}{\partial t} + u\frac{\partial w}{\partial x} + w\frac{\partial w}{\partial z} = -\frac{\partial p}{\partial z} + B + \frac{\partial \tau_{13}}{\partial x} + \frac{\partial \tau_{33}}{\partial z} \tag{3}$$

$$\frac{\partial u}{\partial x} + \frac{\partial w}{\partial z} = 0 \tag{4}$$

$$\frac{\partial B}{\partial t} + u\frac{\partial B}{\partial x} + w\frac{\partial B}{\partial z} = \frac{\partial H_1}{\partial x} + \frac{\partial H_3}{\partial z} \tag{5}$$

The basic geostrophic flow is driven by the background pressure field P_o, which is assumed to have constant gradients in the x- and y- directions. For convenience, temperature, T, has been replaced by buoyancy $B = g(T-\bar{T})/\bar{T}$; where $\bar{T}$ is a reference temperature. The determination of the stress tensor, τ, and buoyancy flux vector, H, is discussed below.

The computational grid is in the (x,z) plane and has a constant spacing in the x-direction, but a stretching scheme is used in the vertical to give increased resolution close to the surface. Periodic boundary conditions are applied in the horizontal and a stress-free condition is applied at the top of the domain. At the lower boundary, profiles of wind and buoyancy are forced to fit standard Monin-Obukhov similarity functions for stably-stratified flow with a constant buoyancy flux applied.

The stress tensor and heat flux vector parametrize the effects of sub-grid scale turbulence and are determined by an eddy viscosity closure, i.e.

$$\tau_{ij} = \nu \left(\frac{\partial u_i}{\partial x_j} + \frac{\partial u_j}{\partial x_i} \right) \tag{6}$$

$$H_i = \nu \frac{\partial B}{\partial x_i} \tag{7}$$

A mixing length hypothesis then determines the eddy viscosity ν:

$$\nu = \ell^2 S \tag{8}$$

where the deformation, S, is defined by:

$$
S = \left\{ \frac{1}{2} \left[\frac{\partial u_i}{\partial x_j} + \frac{\partial u_j}{\partial x_i} \right] \left[\frac{\partial u_i}{\partial x_j} + \frac{\partial u_j}{\partial x_i} \right] \right\}^{1/2}
\tag{9}
$$

To complete the closure, we must now specify the mixing length, ℓ. Under conditions of neutral static stability, and close to the ground, we expect to find $\ell \sim k(z + z_o)$, where k is the von Karman constant and z_o is the roughness length. Away from the ground ℓ will not increase indefinitely but will approach some value ℓ_o. Stable stratification will tend to damp out turbulence and hence reduce the effective value of ℓ. If we define a Richardson number:

$$
Ri = \frac{\partial B}{\partial z} S^{-2}
\tag{10}
$$

then we can construct a stability function to account for this effect. The function is defined by:

$$
\emptyset = \begin{cases} (1 - C_s Ri) & (Ri < C_s^{-1}) \\ \\ 0 & (Ri \geqslant C_s^{-1}) \end{cases}
\tag{11}
$$

For neutral stability (Ri = 0), $\emptyset = 1$, but as some critical Richardson number, C_s^{-1}, is approached, $\emptyset$ approaches zero linearly. A second-order closure turbulence model could be used to generate the functional dependence of $\emptyset$ on Ri. However given the above limiting values of $\emptyset$, its functional dependence on Ri can not be very complicated and the linear approximation used here is probably sufficient. Writing

$$
\frac{1}{\ell} = \frac{1}{(k \emptyset (z + z_o))} + \frac{1}{(\emptyset \ell_o)}
\tag{12}
$$

then produces a mixing length with the required variation with both height and stability. All that is required to complete the specification are values for ℓ_o and C_s. Measurements in neutral boundary layers would suggest that ℓ_0 should be about 40m (Taylor (1969)). In fact the model proves to be fairly insensitive to the exact value of this parameter. The linear theory of the stability of stratified shear flows (see e.g. Chandrasekhar (1961)) shows that such flows are stable for Richardson numbers greater than or equal to 0.25. This would suggest a value of 4 for C_s. However, C_s^{-1} is the Richardson number at which a fully non-linear turbulent flow is completely damped by stability effects; this could well be higher than the value for the onset of linear instability. Mason and Sykes (1982) show that using a value of C_s greater than 3 can give rise to grid-length "roughness" in the model fields, suggesting that the sub-grid turbulence has been inadequately parametrized. $C_s = 3$ was used for most integrations in this study but the effects of varying it are discussed below.

It is expected that horizontal exchanges, such as τ_{11}, and H_1, will not be affected to any great extent by stability, for this reason a horizontal viscosity was defined as:

$$\nu_H = \ell_0^2 \, S_H \tag{13}$$

where S_H is the contribution to the total deformation from horizontal gradients only. Where $\nu_H > \nu$, it was used in the calculation of the horizontal diffusion terms in the equations of motion, (1)-(5).

The numerical scheme for integrating the equations was identical to that used by Mason and Sykes, but a different method was adopted to produce the initial fields used by the model. Equations (1)-(5) were written in one-dimensional form (i.e. no x-dependence) and an iterative method was used to obtain a steady-state solution for a neutral boundary layer (B≡0). A constant positive buoyancy flux, implying cooling, was applied at the lower boundary, and the equations were integrated forward in time for 15000 seconds, at the end of which period a stably-stratified boundary layer profile had developed. Because simple explicit time-stepping schemes were used, the time-step had to be less than 0.25s. The one-dimensional fields of u, v and B thus obtained were then used to initialise the two-dimensional computational domain and the two-dimensional integration proceeded. In order to perturb the

initial fields, the surface stress was doubled over one third
of the domain for the first few tens of time-steps.

RESULTS

   Before carrying out an integration, several parameters
including the magnitude and direction of the geostrophic wind
(fixed by the basic pressure gradients $\frac{\partial P_O}{\partial x}$ and $\frac{\partial P_O}{\partial y}$, and f),
the size of the domain, the horizontal and vertical grid
spacings and the turbulence constants C_s and ℓ_O must be fixed.
In addition the form of the initial fields will be affected by
the buoyancy flux and the length of time over which the one-
dimensional "set-up" model was integrated; these were kept
fixed at $0.001 \text{m}^2 \text{s}^{-3}$ and 15000 s respectively. This buoyancy
flux corresponds to a downward heat flux of about 30 Wm^{-2} which
is typical of the sensible heat fluxes observed in a nocturnal
boundary layer under partially clear skies. For all the runs
described here the geostrophic wind was 10 ms^{-1} in the
x-direction. This is a fairly arbitrary choice since there is
no reason to suppose that the most energetic modes will be
aligned across the geostrophic wind, but this choice does give
the maximum available shear across the boundary layer as a
whole.

   The depth of the domain was fixed at 3 km, which is
sufficient to accommodate a boundary layer which is typically
300m deep while leaving room for pressure field disturbances
to decay before reaching the upper boundary. Initially the
mesh-stretching algorithm was set to give a vertical grid
spacing of about 8m at the ground, increasing to 32m at the
top of the boundary layer and 400m at the top of the domain,
using a total of 40 grid points. A domain width of 300m was
chosen after several trial integrations and 64 evenly-spaced
points were used in this direction. This will be referred to
as the "standard grid".

   The standard choice of turbulence parametrization constants
has been discussed in the previous section.

(a)   Integration with standard grid and standard turbulence
      constants

   The fields of u, v and B produced by the one-dimensional
model, after "cooling" with buoyancy flux $\overline{wB} = 10^{-3} \text{m}^2 \text{s}^{-3}$ for
15000 seconds, are shown in Fig. 1. Despite the rather crude

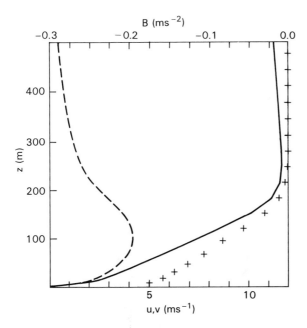

Fig. 1   Profiles of the mean velocity components u (———)
         along the computational domain, across the computa-
         tional domain v (-----) and buoyancy B = g(T-T̄)/T̄
         (+ + + +) obtained from integration of a one-
         dimensional model and used as initital conditions for
         the standard run.   The model domain extends up to 3 km;
         only the lowest 500m of the profiles are shown here.

turbulence parametrization, the profiles show similar features
to those observed in the nocturnal boundary layer, in
particular u has a maximum, or weak "nocturnal jet" at 200m.
Holding the geostrophic wind speed and surface heat flux
constant simplifies the evolution of the profiles; in the
atmospheric boundary layer these ideal steady conditions are
rarely encountered.

    The two-dimensional integration was started with these
profiles as initial conditions.   Figs. 2a, b ·c show the evolution of the vertical velocity field in the lowest 500m of the
computational domain.   After 500 seconds (2a), the initial
perturbation has decayed and a rather short wavelength mode
(λ = 21m) has become established near the top of the stable
layer.   At 1250 seconds, this mode is still present but appears
at a slightly longer wavelength.   In addition a second mode of
wavelength 150m has appeared, and is apparently trapped within

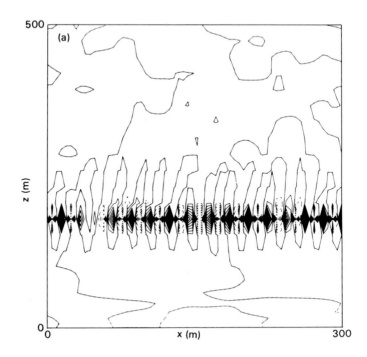

Fig. 2   Contours of the vertical velocity field in the lowest
         500m of the 300m wide two-dimensional model domain for
         a run with "standard" turbulence parametrization and
         resolution (see text). The contour interval is one-
         tenth of the maximum value of vertical velocity, broken
         contours denote negative vertical velocity.

         (a)   Field after 500s from the start of the
               integration, w_{max} = 5 mms^{-1}.

the stable layer. After 6000s both modes are still present.
The short-wavelength mode now has a wavelength of 37.5m while
the trapped mode has expanded to wavenumber 1 (λ = 300m) and
is probably influenced by the limited horizontal extent of the
domain. By this stage the total eddy kinetic energy in the
domain had reached a steady value, and the integration was
terminated. Throughout the integration, the stratification

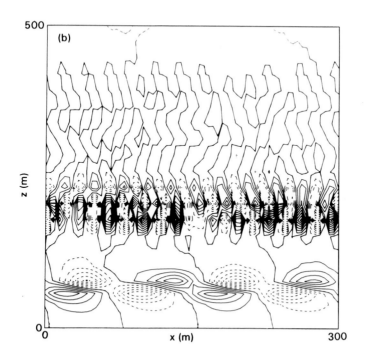

Fig. 2b Field after 1250s, w_{max} = 102 mms^{-1}.

remained stable at all points in the domain, indicating that
there was no tendency for the waves to break.

   Turbulent transports in the model are due partly to the
small-scale turbulence parametrization, and partly to any
eddies which may be resolved within the domain. Fig. 3 shows
vertical profiles of the contributions to the vertical
transport of horizontal momentum from the turbulence parametri-

zation (i.e. $\nu \frac{\partial u}{\partial z}$) and from resolved eddies, $- \overline{u'w'}$.   These

quantities have been averaged horizontally over the domain and
also in time for the latter part of the integration. The
stress due to the resolved eddies is everywhere much smaller
than that due to the turbulence parametrization and it is
significant only in a very thin layer around the level of
maximum amplitude of the long-wavelength trapped mode.

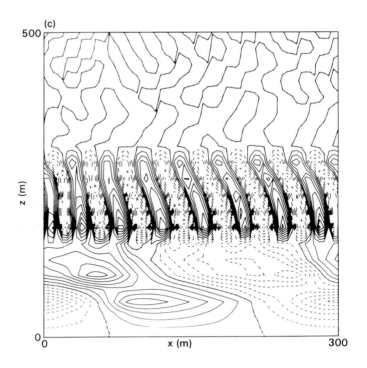

Fig. 2c   Field after 6000s, $w_{max} = 210 mms^{-1}$.

Because the integration did not reach a steady state while
statistics were being accumulated, this maximum of resolved
eddy stress may have been somewhat smeared out.

Fig. 4 shows corresponding profiles for buoyancy flux.
The contribution from the resolved eddies is seen to be
comparable with the parametrized part in a very limited region.
The parametrized part appears to be somewhat reduced in this
same region in partial compensation for the resolved-scale
transport.

The absence of wave breaking is reflected in these low
values of wave-induced heat and momentum fluxes, which imply
that the waves exert only a small influence on the evolution
of the mean profiles.

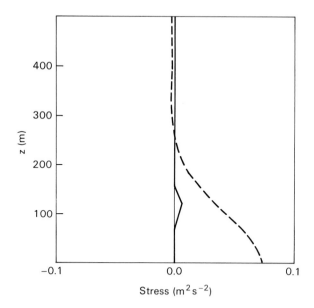

Fig. 3   Profiles of resolved eddy stress, $-\overline{u'w'}$, (solid line)

and parametrized stress, $\nu\dfrac{\overline{\partial u}}{\partial z}$, (dashed line) in the
standard integration, averaged over the period from
2500s to 5000s after the start of the integration.

(b)   "Quasi-Linear" run

In order to identify the most rapidly growing mode, it is
possible to conduct a "quasi-linear" integration of the
equations.   In this case the integration initially proceeds
normally until the root mean square value of vertical velocity
(averaged over the computational domain) reaches a value of
0.001 ms^{-1}.   The deviations of u, v, w and B from their
original profiles are then reduced by a factor of 10 and the
integration proceeds until $w_{RMS} > 0.001$ ms^{-1} once more; the
process is then repeated.   By keeping the perturbations to the
initial fields very small, the nonlinear terms in (1)-(5) will
be negligible, and only the fastest growing linear modes will
actually appear in the integration.

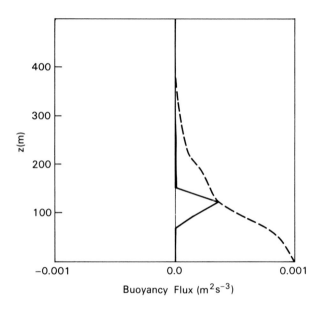

Fig. 4   Profiles of resolved eddy buoyancy flux, $\overline{wB}$, (solid

line) and parametrized buoyancy flux, $\nu \dfrac{\partial \overline{B}}{\partial z}$, (dashed

line) in the standard integration, averaged over the
period from 2500s to 5000s after the start of the
integration.

Fig. 5 shows the vertical velocity field from a quasi-linear
integration with all parameters as in the standard run
described previously. The only mode present corresponds to
the short wavelength instability observed in the non-linear
integration described above; it has a wavelength of 18.75m.

(c)   Interpretation

The short wavelength mode at the top of the boundary layer
is possibly the simpler of the two to understand. The weak
jet in the u-profile gives rise to an inflection point leading
to the possibility of classical Kelvin-Helmholtz instability.
The quasi-linear integration and the initial stages of the non-
linear integration show that the most rapidly growing modes of
this instability have quite short wavelengths ($\leqslant$ 20m), but

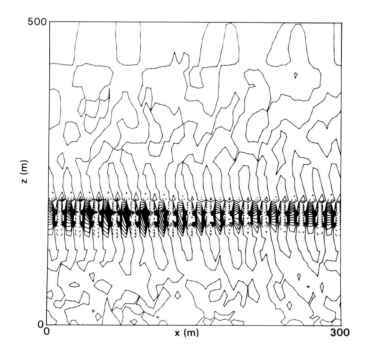

Fig. 5 Contours of the vertical velocity field after 1250s of
        a "quasi-linear" integration. The area shown is as in
        Fig. 2.

non-linear effects appear to damp out these shorter scales.
Examination of the resolved eddy flux profiles shows that this
mode gives a negligible contribution to transport processes.

    The long wavelength trapped mode is of greater interest.
Davis and Peltier (1976) have shown the existence of instabilities in a stratified shear flow which does not have an inflection point in the velocity profile, and thus cannot support
Kelvin-Helmholtz instability. This mode grows by a process of
"resonant overreflection"; it is trapped between the ground
and a region of low Richardson number (in the present case, at
the top of the boundary layer).

    The structure of the long wavelength mode in the current
model is very similar to that found by Davis and Peltier (1979)

in a linear stability analysis of profiles similar to those
used in this study. In both cases the mode has a minimum of
amplitude (see Figs. 2b, 2c) at a critical level within the
boundary layer, where the phase speed of the wave is equal to
the local flow speed. Davis and Peltier (1979) also present a
profile of eddy stress and show that this is everywhere small
except around the critical level. This corresponds well to the
peak in resolved eddy stress seen in Fig. 3. There are thus
good grounds for identifying the long wavelength trapped mode
with the resonant overreflection class of instabilities.

Davis and Peltier have calculated linear growth rates for
the resonant overreflection instabilities and have shown them
to be small when compared with typical Kelvin-Helmholtz growth
rates. However, they have also demonstrated that if Kelvin-
Helmholtz modes are allowed to interact nonlinearly with
resonant overreflection modes, the latter can reach a significant amplitude much faster than linear growth alone would
permit. This explains the evolution of the modes observed in
the present work. The quasi-linear integration demonstrates
that the short wavelength (Kelvin-Helmholtz) mode has the
highest linear growth rate and that the long wavelength mode
is either nearly stable or has a very small growth rate. In
the non-linear integration, however, the long wavelength
(resonant overreflection) mode appears to grow quite rapidly,
but only after the short wavelength mode has reached finite
amplitude.

(d)   Variation of Turbulence Parametrization

Altering the values of the turbulence constants C_s and ℓ_o
will affect both the form of the undisturbed boundary layer
and the development of eddies. The results of some trial
integrations suggest that variation of ℓ_o by 50% either way
has only a small effect. This is because the stability
function ϕ in (12) is the dominant factor in limiting the
mixing length away from the ground. Obviously under neutral
or near neutral conditions the role of ℓ_o will become much more
important.

Integrations with $C_s = 2$ (i.e. critical Richardson number =
0.5) start with an unrealistically deep boundary layer. The
types of eddy which develop are similar to those observed in
the standard run but are considerably less energetic. There is
little theoretical justification for using such a high value
of the critical Richardson number and the unrealistic initial
fields would support the choice of a lower value.

When values of C_s greater than 3 are used the eddies are
very energetic and develop rapidly. Resolved eddy fluxes are
much greater than those in the standard integration and cause
significant changes in the boundary layer structure during the
run. However, inspection of the fields reveals a certain
amount of "roughness" on scales of one or two grid lengths,
which would suggest that the parametrization of small-scale
turbulence is inadequate. These runs in which C_s was varied
lend support to the original choice of C_s = 3.

(e) Variation of Resolution

In order to test the sensitivity of the model to vertical
resolution, the vertical grid stretching algorithm was changed
to decrease the vertical mesh spacing within the boundary
layer to about 9m at the expense of the resolution above the
boundary layer. This change in model resolution produces some
dramatic changes in the form of the observed modes.

Fig. 6a shows the vertical velocity field after 1000
seconds. The only mode present is an extremely fine scale
disturbance trapped within the boundary layer. The fields have
not been spectrally analysed, but it is clear that there is a
significant amount of energy present on scales of 1 - 4 grid
lengths, which are only poorly resolved. The disturbance is
quite weak, with maximum vertical velocity of only 30 mms^{-1}.
The Kelvin-Helmholtz mode is completely absent at this stage;
this may be due to the reduced resolution at the top of the
boundary layer leading to a filtering out of the shortest
wavelength (and fastest growing) modes of this instability.

After 5000 seconds (Fig. 6b) the Kelvin-Helmholtz mode has
appeared. The very short wavelength features in the trapped
mode appear to have been damped out and the remaining disturbance has a wavelength of about 43m - very much shorter than
the 300m wavelength mode observed at this stage in the integration on the standard grid. Peak vertical velocities are still
only 61 mms^{-1}, compared with 210 mms^{-1} in the standard run and
the resolved eddy fluxes of heat and momentum are negligible.
The Kelvin-Helmholtz instability has a wavelength of 23m
compared with 37.5m at a similar stage in the standard integration. This suggests that nonlinear interaction between the
short wavelength trapped mode and Kelvin-Helmholtz instability
may be leading to shorter wavelength modes of the latter being
selected.

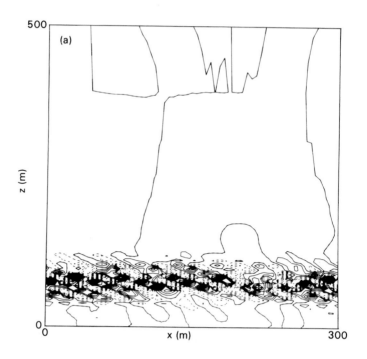

Fig. 6   Contours of the vertical velocity field for an integra-
         tion with standard turbulence parametrization but with
         increased vertical resolution within the boundary
         layer.   The area shown is as in Fig. 2.

         (a)   Field after 1000s, $w_{max} = 29$ mms^{-1}.

     The short wavelength trapped mode has a structure resembling
a resonant overreflection instability.   However, Davis and
Peltier (1979) did not find any modes on such short length
scales in their stability analysis of a similar system.   It is
not entirely surprising that the increase in resolution leads
to the appearance of shorter wavelength modes; what is somewhat disturbing is the complete suppression of the long wavelength trapped mode.

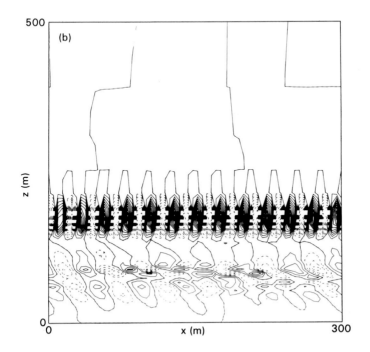

Fig. 6b Field after 5000s, $w_{max} = 61$ mms^{-1}

DISCUSSION

Integrations of a model of the stable atmospheric boundary
layer have revealed at least two modes of two-dimensional
instability. These have been tentatively identified as a
Kelvin-Helmholtz instability associated with the weak jet
profile at the top of the boundary layer and a resonant overreflection instability trapped within the boundary layer.

Neither mode makes a significant contribution to the fluxes
of heat and momentum (except possibly in the region of the
critical level of the resonant overreflection mode) unless an
unrealistic parametrization of small-scale turbulence is used.
This should not be taken to mean that large scale eddy motions
are unimportant in the stable boundary layer; it is possible
that three-dimensional motions or eddies aligned other than at

right angles to the geostrophic wind could generate significant fluxes. The former are outside the scope of the present study. A limited number of integrations have been conducted with the geostrophic wind at various angles to the computational domain without revealing any new features but the angular coverage was far from complete.

Given the above reservations about the possible existence of other modes, the absence of energetic eddies suggests that the turbulence parametrization is sufficient to describe the fluxes. This gives some encouragement to the application of simple mixing length turbulence models to the stable boundary layer.

A three-dimensional "large-eddy" model of the stable boundary layer would reveal whether three-dimensional modes were important. However, such a model should also support the two-dimensional modes investigated here and this study may prove helpful in interpreting the results from three-dimensional models when they are developed.

The dependence of the scale of the observed eddies on vertical resolution is difficult to explain. The linear growth rates of the resonant overreflection instabilities are small and their evolution is governed by nonlinear interactions which must be quite weak given the small amplitudes observed. It would appear that the short wavelength trapped mode observed with fine vertical resolution must act nonlinearly to suppress development of longer wavelength overreflection instabilities.

ACKNOWLEDGEMENTS

I would like to thank Dr. P.J. Mason for helpful discussion and Mr. J.S. Burgess for carrying out some preliminary investigations.

REFERENCES

Caughey, S.J. and Readings, C.J. (1975) An observation of waves and turbulence in the Earth's boundary layer. *Bound. Layer Met.*, **9**, 279-286.

Chandrasekhar, S. (1961) Hydrodynamic and Hydromagnetic Stability. Oxford: Clarendon Press.

Davis, P.A. and Peltier, W.R. (1976) Resonant parallel shear instability in the stably stratified planetary boundary layer. *J. Atmos. Sci.*, **33**, 1287-1300.

Davis, P.A. and Peltier, W.R. (1979) Some characteristics of
    the Kelvin-Helmholtz and resonant overreflection modes of
    shearflow instability and of their interaction through
    vortex pairing. *J. Atmos. Sci.*, **26**, 2394-2412.

Hunt, J.C.R., Kaimal, J.C., Gaynor, J.E. and Korrell, A. (1983)
    Observations of turbulence structure in stable layers at the
    Boulder Atmospheric Observatory. In "Studies of Nocturnal
    Stable Layers at B.A.O.", (J.C. Kaimal, ed.), Boulder:
    U.S. Dept. of Commerce.

Mason, P.J. and Sykes, R.I. (1982) A two-dimensional numerical
    study of horizontal roll vortices in an inversion capped
    planetary boundary layer. *Quart. J.R. Met. Soc.*, **108**,
    801-823.

Merrill, J.T. (1977) Observational and theoretical study of
    shear instability in the air flow near the ground. *J.
    Atmos. Sci.*, **34**, 911-921.

Taylor, P.A. (1969) On planetary boundary-layer flow under
    conditions of neutral thermal stability. *J. Atmos. Sci.*,
    26, 427-431.

STRATIFICATION AND INTERNAL WAVES IN THE WESTERN IRISH SEA

M.F. Lavin

(Department of Physical Oceanography, Marine Science Laboratories, University College of North Wales)

and

T.J. Sherwin

(Unit for Coastal and Estuarine Studies, Marine Science Laboratories, University College of North Wales)

INTRODUCTION

Whilst most of the Irish Sea remains well mixed throughout the year, the area to the south-west of the Isle of Man becomes thermally stratified in the summer months (Fig. 1). The stratified region is bounded to the south and east by a region of strong horizontal temperature gradients, or front, the position of which is determined by a balance of surface heating and tide and wind stirring. Associated with this front are eddies, residual currents, local convergences and divergences and high biological activity. The work presented here forms part of an extensive programme of observations designed to improve our understanding of the seasonal development and internal dynamics of the stratified region. At Station B8 (Fig. 1), temperature was recorded every 10 minutes at a series of depths between 7th April and 13th October 1981. In addition, currents were measured 17 m below the surface. This resumé is in two parts: the first covers the development of stratification, from onset in Spring to breakdown in Autumn; and the second summarises the nature of the internal wave regime during the month of August.

THE SEASONAL CYCLE

The western Irish Sea usually presents some degree of stratification from early April, the extent of which is very weather dependent. In 1981, there was some stratification before the start of the experiment, but it was completely mixed by a large storm on April 24th to 26th. Stratification recommenced on May 1st and was firmly established by May 10th (Fig. 2). The thermocline, which did not deepen during the

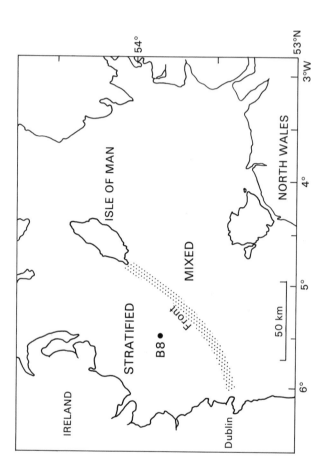

Fig. 1   The summer stratification of the Irish Sea

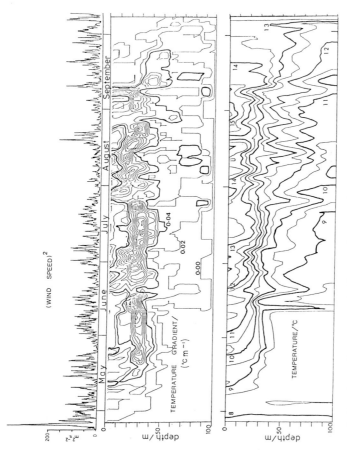

Fig. 2 The square of the hourly mean wind speed at Dublin airport, and the low-passed (zero filter response at 2 days) time series of the vertical temperature distribution at Station B8 during 1981. The temperature gradient is contoured every 0.02 °C m^{-1}. The 0.04 °C m^{-1} contour delimits the thermocline.

summer, was centred on about 25 m. A large degree of variability with periods longer than two days is apparent at all
depths. Most of the variability near the surface was due to
meteorological factors; there is a clear visual correlation
between wind stress and the depth of the surface mixed layer.
Some of the variability in the deeper layers may be due to
horizontal advection, and some of it coincides with periods of
high internal wave activity. The period of stratification
breakdown began on about September 12th, with convective
mixing due to surface cooling becoming important. On October
6th, warm (13.5 $^\circ$C) mixed water advected to Station B8. After
the breakdown phase, the water remains well mixed throughout
the winter, gradually cooling until it reaches a minimum of
about 7.5 $^\circ$C in February.

The seasonal cycle can be effectively summarised by
hysteresis loops (Fig. 3), which are also used to assess the
performance of numerical models (A.E. Gill and J.S. Turner
(1976)). The heat content (Q) and potential energy anomaly (Φ)
of the water column are defined by

$$\Phi = (1/h) \int_{-h}^{0} (\bar{\rho}-\rho) zg \ dz$$

$$Q = C_p \bar{\rho} \int_{-h}^{0} T(z) \ dz;$$

where

$$\bar{\rho} = (1/h) \int_{-h}^{0} \rho(z) \ dz;$$

$T(z)$ and $\rho(z)$ are temperature and potential density at depth
z; h is the total depth of water (100 m); and C_p is the
specific heat of sea water. Φ is the amount of mechanical
energy required to completely mix the water column. The maximum surface temperature ($T_s = 15.25 \ ^\circ$C) was recorded on August
29th and the maximum heat content ($Q = 53.15$ J m^{-3}) on
September 14th to 22nd (ignoring the advective event on October
6th). While T_s and Q show definite maxima, Φ remains around

75 J m^{-3} from mid-July to the beginning of September. It takes
longer for Φ to reach this plateau (from April to mid-July)
than it does for the stratification to be destroyed (from early
September to mid-October).

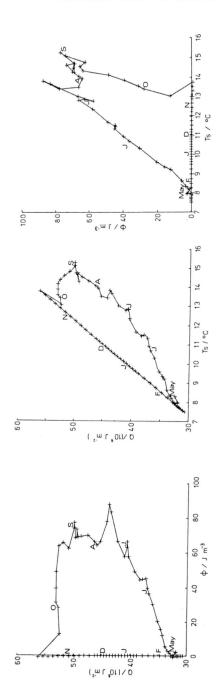

Fig. 3 Hysteresis loops. The data from Station B8 were low-passed (zero filter response at 8 days), and the annual cycle completed by interpolation of data collected in the region of Station B8, from a variety of sources. T_s is the sea surface temperature; Q and Φ are the heat content and potential energy anomaly of the water column respectively. The beginning of each month is marked with its initial letter and there are four days between crosses.

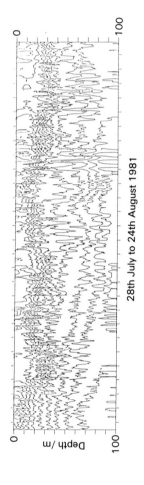

Fig. 4   Isotherm displacements. The data have been block averaged over 30 minute intervals
and contoured every 0.5 °C. The highest temperature contour is 15 °C and the lowest
is 9.5 °C. Note the occasional appearance of large amplitude oscillations with a semi-
diurnal period – these suggest the presence of a variable internal tide.

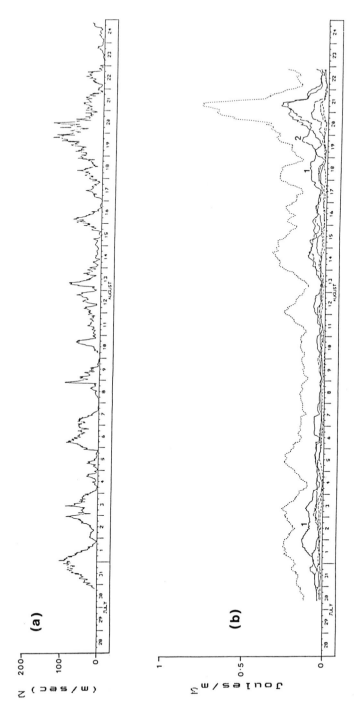

Fig. 5 (a) Time series of the square of the wind speed at Station B8. (b) Potential energy of the first five modes of waves with periods ranging from 40 minutes to 9 hours - modes 1 and 2 dominate. The dotted line is the total potential energy of all modes. These data seem to support the idea that the wind can generate internal waves, apparently with a lag of the order of a day. The energy is probably transmitted from the wind via inertial currents and large scale baroclinic responses.

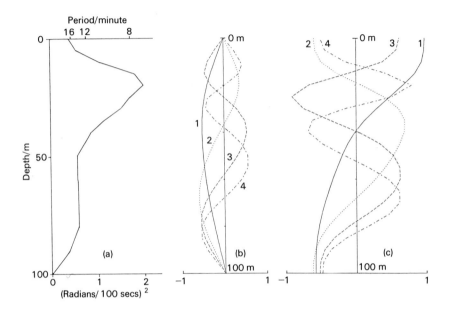

Fig. 6   Buoyancy frequency and modal profiles.

(a) $N(z)^2 = g/\rho *d\rho/dz$ (where N is the local buoyancy
frequency) averaged over time (one month) and space
(10 m). Waves with periods greater than 16 minutes
are free to propagate through the water column, but
those with shorter periods are contained within the
pycnocline.   (b)   The theoretical variation in vertical
displacement of the first 4 modes of a wave with a two
hour period. Waves with longer periods have an almost
identical shape.   (c)   The horizontal displacements
corresponding to (b).   It is apparent that the thermocline is a region of large internal shear.   As the
inertial period (14.86 hrs.) is approached, Coriolis
acceleration causes the horizontal component to rotate,
and the potential energy/kinetic energy ratio to tend
to zero.   In (b) and (c) amplitudes are in arbitrary
units.

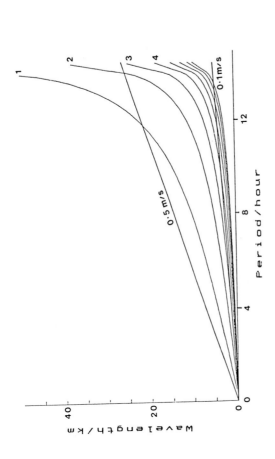

Fig. 7 Dispersion relations. Individual modes are strongly dispersive at low frequencies, where Coriolis acceleration is significant.

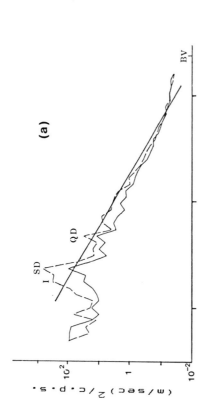

(a)   The variance of horizontal currents at 17 m depth – the broken and plain
      lines are clockwise and anti-clockwise variance respectively.

Fig. 8   Raw Energy Spectra.

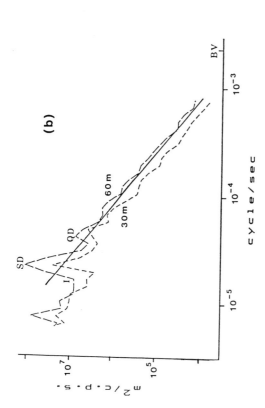

8(b) The variance of vertical displacement at 30 m and 60 m. The spectra have been logarithmically averaged with degrees of freedom ranging from 6 at low frequencies to 600 at high frequencies. The straight lines have the slope of the deep-ocean empirical relationship energy density $\propto$ (frequency)$^{-2}$ (C.J.R. Garrett and W. Munk, 1972). Although in (a) the barotropic tidal currents have been removed by harmonic analysis, there still remains significant clockwise energy at the tidal frequency (SD). The inertial frequency (I) has little vertical energy, but a large amount of clockwise rotation. Quarter diurnal peaks (QD) are detectable and the buoyancy cutoff, or Brunt-Vaisala frequency N is shown as BV on the abscissa.

INTERNAL WAVES

The stratified layers of the western Irish Sea are subject
to continual wavelike oscillations (Fig. 4) which may have an
important influence on the dynamics of the region. This
section summarises the internal wave activity during August
1981. Figs. 4 and 5 are the time series of temperature structure, wind stress and internal wave potential energy. Figs.
6 to 8 show the average values of relevant internal wave parameters. For a recent review of internal wave theory see e.g.
W. Munk (1981). A fuller report has been prepared by Sherwin,
T.J. (1983).

ACKNOWLEDGEMENTS

This work was partly supported by a grant from the Natural
Environmental Research Council. CONACyT (Mexico) provided a
studentship for MFL.

REFERENCES

Garrett, C.J.R. and Munk, W. (1972) *Geophysical Fluid
    Dynamics*, **2**, 225-264.

Gill, A.E. and Turner, J.S. (1976) *Deep Sea Research*, **23**,
    391-401.

Munk, W. (1981) Evolution of Physical Oceanography. MIT Press
    Press, 292-341.

Sherwin, T.J. (1983) Internal waves in the Irish Sea. UCES
    Paper U83-3, Marine Science Lab., Menai Bridge, Gwynedd.

THE BENTHIC BOUNDARY LAYER

K.J. Richards

(Institute of Oceanographic Sciences, Wormley)

ABSTRACT

The benthic, or bottom, boundary layer of the ocean plays an important role in the structure and dynamics of deep oceanic flows. Not only is it an important sink of energy but it can have a significant effect on the interior motions of the ocean. This paper reports on mathematical models of the vertical structure of the boundary layer and its horizontal variability. Firstly a second order closure model is used similar to that used in the stable atmospheric boundary layer, but unlike its atmospheric counterpart the benthic boundary layer is also controlled by its erosion of the stable stratification above it and by the upward propagation of internal waves. Another special feature of these boundary layers is that the motion of large mesoscale eddies moving in the stable flow above the boundary layer can completely change the structure of flow within the boundary layer and in fact determine its height.

1. INTRODUCTION

The rotation of the Earth has a strong influence on currents within the ocean. Away from boundaries and regions of strong currents the flow is to a good approximation geostrophic and follows pressure contours. Close to the ocean floor the flow is retarded allowing some flow to cross pressure contours. Because of vertical shear the flow becomes turbulent. The turbulence mixes properties such as temperature and salinity to produce a homogeneous layer some tens of metres thick. The layer is sometimes capped by a region of strong density gradient inhibiting exchange of properties between the mixed layer and above. Turbulence generated at the bottom is restricted to this layer. This layer is referred to as the benthic boundary layer. The term benthic is derived from the

Greek word benthos meaning deep ocean.

The benthic boundary layer is considered to be an important
sink of energy for motions of the ocean interior through the
action of Ekman pumping. It is often the major sink of energy
in numerical models of the ocean (e.g. Holland, 1978). Recent
interest in this region of the ocean stems from assessment
studies of the feasibility of the disposal of heat generating
radioactive waste into the deep sea environment. A knowledge
of the dynamics of the flow is required to understand the
initial mechanisms of dispersal of radionuclides introduced
into the benthic boundary layer.

A distinction must be made between the height of the bottom
mixed layer and the height to which the flow is affected by the
presence of the boundary, the momentum or Ekman layer. In
cases where the flow varies horizontally the height of the
mixed layer is distorted by Ekman pumping and may be many times
thicker than the height of the Ekman layer.

Until recently observations of the benthic boundary layer
have been limited. Measurements of temperature and salinity
showed a mixed layer often several times that expected by
existing one dimensional models (see e.g. Armi and Millard,
1976). Armi and D'Asaro (1980) studied the bottom mixed layer
at a site in the W. Atlantic and found the height of the mixed
layer to vary between 5 and 60 m over horizontal distances of
20 km and periods of 15 days. The layer is at least
intermittently turbulent (D'Asaro, 1982) with the turbulence
being confined to the mixed layer. The much weaker
stratification in the E. Atlantic (Saunders, 1983) and E.
Pacific (Hayes, 1979) makes it difficult to detect the presence
of a mixed layer. However, by a careful analysis of
temperature measurements made in the E. Atlantic experiment it
has been possible to deduce that the average height of the
layer is approximately 30 m, becoming less than 10 m on some
occasions and greater than 100 m on others (Saunders, private
communication). A striking feature of the temperature records
is the presence of fronts of about 3 m$^{\circ}$C which are persistent
and are advected with the flow (Thorpe, 1983). Recent
measurements of the flow above the bottom carried out by IOS
using electromagnetic current meters show the flow to be
turbulent with the ratio of the variance of the flow to the
mean to be similar to other geophysical boundary layers
(Elliott, 1984).

Two approaches have been taken to model the dynamics of the
benthic boundary layer and are reported on in the next two
sections. Most authors have investigated the vertical
structure of the layer by using a one-dimensional model of the

region and assuming the flow to be horizontally uniform. The
models provide estimates of the expected height of the layer
and how long it takes to reach this height together with the
vertical variations of the flow and turbulence quantities.

The observations of Saunders (1983) show that the height of
the mixed layer is not correlated with the strength of the flow
above the layer as is implied in a one-dimensional model. A
second approach has therefore been taken to try and account for
this discrepancy between theory and observations. Horizontal
variations in the strength and direction of the current are
taken into account. These variations cause convergences and
divergencies of the flow within the mixed layer. The mixed
layer is thickened and thinned, respectively, a process called
Ekman pumping.

A major contribution to the horizontal variation in currents
in the deep ocean is from eddies which have diameters 50-200 km
and speeds a few centimetres per second. They are commonly
referred to as mesoscale eddies. Over the Madeira abyssal
plain in the E. Atlantic mesoscale eddies were found to have a
diameter of approximately 40 km and an average speed of 1 to
2 cms^{-1} (Saunders, 1983). A numerical model has been developed
to study their effect on the bottom mixed layer (Richards,
1984). The model provides estimates of variations in both height
and temperature of the mixed layer. It also gives an estimate
of the residence time for particles to remain within the mixed
layer.

2. VERTICAL STRUCTURE

The current close to the sea bed has three main components,
the linear (semi-diurnal) tide, inertial oscillations and a
slowly varying flow resulting from mesoscale eddies. On the
Madeira abyssal plain these have periods of 12.4 hours, 22
hours and 50 to 100 days respectively. Because of the
relatively short development time of the mixed layer (a few
days) the time variation of the mesoscale motions can be
ignored when discussing the vertical structure of the bottom
layer. The unsteadiness of the flow at the tidal and inertial
frequencies, however, does have to be taken into account.

The height of the boundary layer depends upon the Earth's
rotation, the distribution of density and the unsteadiness of
the flow. We consider a horizontally homogeneous flow above a
horizontal plane surface. The equations for the mean velocity
components, U, V and mean potential temperature θ are

$$\frac{\partial U}{\partial t} + \frac{\partial}{\partial z} \overline{uw} = f(V - Vg) \qquad (2.1)$$

$$\frac{\partial V}{\partial t} + \frac{\partial}{\partial z} \overline{vw} = f(Ug - U) \qquad (2.2)$$

$$\frac{\partial \theta}{\partial t} + \frac{\partial}{\partial z} \overline{w\theta'} = 0 \qquad (2.3)$$

where the coordinate z is taken vertically upwards. Ug, Vg are the components of the geostrophic flow representing the pressure force, u, v, w and θ' the flunctuating components of velocity and temperature, respectively and f the Coriolis parameter. The density stratification is characterised by the buoyancy frequency $N = [-g/\rho \frac{\partial \rho}{\partial z}]^{\frac{1}{2}}$ where g is the acceleration due to gravity and ρ the density of the fluid. Typical values of N are 7×10^{-4} s^{-1} for the area studied by Armi and D'Asaro (1980) in the W. Atlantic and approximately 2×10^{-4} s^{-1} or less for the area studied by Saunders (1983) in the E. Atlantic. The initial state is taken to be a linear density gradient, i.e. constant N. It is also assumed that there is no net heat flux through the region considered. Wimbush and Munk (1970) have shown that the flux of heat due to geothermal heating is negligible. Any flux of cold water coupled to the general circulation of the ocean has also been neglected. Little is known about this flux but it is unlikely to affect the development of the bottom mixed layer which occurs on a relatively short timescale.

Equations 2.1 - 2.3 need to be closed by specifying the Reynolds stress terms. If a constant eddy viscosity is used the classical Ekman spiral is produced with the flow at the surface being 45° to the left of the flow above. With a closure scheme that gives a logarithmic layer at the bottom this is reduced to approximately 10° - 20°. Two studies have applied higher order closure schemes to the problem. Weatherly and Martin (1978) use the 'Level II' model of Mellor and Yamada (1974), a turbulence closure in which the advection and diffusive terms in the Reynolds stress equations are neglected. The turbulent fluxes can be written in eddy coefficient form with the effective eddy viscosity proportional to the square root of the local value of the turbulent kinetic energy and a mixing length dependent upon the distance from the lower boundary and the local stability. Such an approach assumes that the turbulence is in local equilibrium and that the

Reynolds stresses are dependent upon the local shear. A
critical Richardson number is introduced above which turbulent
mixing ceases. They find the height of the developed mixed
layer, h_o , to be given approximately by the expression

$$h_o = 1.3u_*/(f(1+N^2/f^2)^{\frac{1}{4}}) \qquad (2.4)$$

where u_* is the friction velocity. Weatherley and Martin
(1978) have also considered the effects of a sloping bottom and
Weatherley, Blumsack and Bird (1980) the effect of a diurnal
tidal current.

   Richards (1982a) uses a second order closure scheme based on
that used by Wyngaard (1975) for a stably stratified
atmospheric boundary layer. Transport equations are used to
determine the time evolution of individual Reynolds stresses
which are not assumed to be proportional to the local shear.
The change in the turbulence structure (so called rapid
distortion effects) that occur in the time dependent boundary
layer are calculated through the pressure velocity
correlations in the transport equations. The results for a
typical case with the $N = 10^{-3}$ s^{-1}, $f = 10^{-4}$ s^{-1}, and a steady
current of $U_o = 5$ cm s^{-1} are shown in Fig. 1.

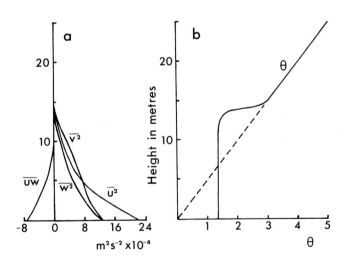

Fig. 1 Profiles of the turbulent stress components $\overline{uw}$, the
variance components $\overline{u^2}$, $\overline{v^2}$ and $\overline{w^2}$ and the mean (non-
dimensional) temperature θ for a stably stratified
steady flow of 5 cms^{-1} with the buoyancy frequency $N =
10^{-3}$ s^{-1} (after Richards, 1982a).

The time is 3 days from the start of the experiment. A mixed
layer of height 13 m has developed which is capped by a region
of strong density gradient. The height of the profiles of the
Reynolds stresses are restricted to this layer with $\overline{w^2}$ falling
off more quickly with height than $\overline{u^2}$ and $\overline{v^2}$ as the top of the
region is approached. The friction velocity $u_* = 4.38 \times 10^{-2} U_o$
and the angle of the flow at the surface to the geostrophic
flow is $13.2°$. As the layer thickens the density gradient
strengthens and inhibits further mixing. This reduces the
growth rate of the height of the layer. The height of the
mixed layer is plotted in Fig. 2 as a function of time for the
case considered above. Also shown is the result for a semi-
diurnal tidal current with a maximum velocity of 5 cms^{-1}. The
mixed layer grows more slowly under a tidal current than a
steady current. The height of the layer for the tidal case is
approximately 25% less than that of the steady case.

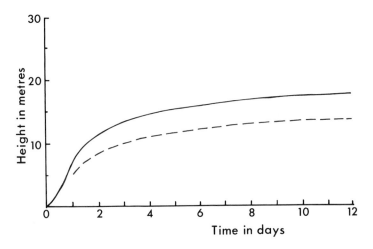

Fig. 2   Plot of the mixed layer height for a stably stratified
         steady (solid curve) and tidal (dashed curve) flow as a
         function of time with the buoyancy frequency $N = 10^{-3}$
         s^{-1} (after Richards, 1982a.)

As the height of the mixed layer increases the local
gradient Richardson number, R_i, at the top of the layer
increases and at some critical height local shear production of
turbulence will be severely inhibited or cease altogether.

In the present case $R_i = 0.25$ when $h \simeq 8$ m. However, in the
model, diffusion of turbulence enables the layer to continue
growing past this height. In fact because there is no net heat

flux in the problem the layer continues to grow indefinitely, albeit at a slow rate. The results of the model of Richards (1982a) suggest that the maximum height of the mixed layer is given approximately by

$$h_o = 0.1 \ U_o (\frac{f}{N})^{\frac{1}{2}}/f \qquad (2.5)$$

The growth of the layer after this height predicted by the model is very small. It is also after this height that internal waves efficiently radiate energy away that was formerly available for increasing the layer depth (Richards, 1982a, see also Carruthers and Hunt (q.v.)) It is noted that the mixed layer height given by (2.5) is approximately twice that given by (2.4) which is based on a criterion based on the local Richardson number. The final stages of development of the model of Richards (1982a) are very dependent on the modelling of the third order correlation terms in the Reynolds stress equations. In particular the diffusion of turbulence is inhibited by stable stratification. Expressions (2.4) and (2.5) should perhaps therefore be treated as lower and upper bounds, respectively, on the equilibrium height of the mixed layer.

Using the results from a number of runs of the model of Richards (1982a) the time development of the height of the mixed layer is given approximately by

$$h = h_o \ (1-e^{-at}) \qquad (2.6)$$

where $a^{-1} \simeq 3$ days. This is approximately twice the development time given by the model of Weatherly and Martin (1978). The reason for this difference is the inclusion of rapid distortion terms in the model of Richards so that the local shear and stress terms are no longer exactly in phase and the production of turbulent kinetic energy is reduced.

Comparing one dimensional models of the benthic boundary layer with observations is difficult principally due to horizontal non-uniformities in the flow (see next Section). Bird and Weatherly (1982) report on velocity and temperature measurements taken on the Eastward Scarp of the Bermuda Rise. They found a steady southward flow following the isobaths. They used the model of Weatherly and Martin (1978) to try to simulate the data. Unfortunately, they were only able to obtain lower and upper bounds on the height of the observed mixed layer of 12 m and 62 m respectively. Expression (2.4) gives $h_o = 41$ m whilst (2.5) gives $h_o = 79$ m. The observed veering of the current direction with height was one half that predicted by the model and the change of mixed layer temperature with time was

predicted to be the wrong sign. As mentioned in the
Introduction, one-dimensional models are unable to predict the
height of the bottom mixed layer in horizontally varying
currents. More measurements are required to test if they can
model the vertical structure within the layer.

3. INTERACTION OF MESOSCALE EDDIES WITH THE BOTTOM MIXED LAYER

   Horizontal variations in the interior flow of the ocean will
cause deformations of the height of the bottom mixed layer
through Ekman pumping. Arguments based on the conservation of
potential vorticity show that a relatively small distortion of
the mixed layer height (O(10m)) will induce an O(1) change in
the vorticity of a mesoscale eddy in contact with the bottom.
A model to study the mesoscale forcing of the benthic boundary
layer needs to be coupled with a model of the ocean interior.
Richards (1984) has developed a quasi-geostrophic model (i.e.
the flow is taken to be close to geostrophic) to study this.

   The flow is considered in an open box 500 km square which is
assumed to be periodic in both horizontal directions (see
Fig. 3). In the vertical the model has three layers. The
upper two layers model the ocean interior. Their depths H_1 and
H_2, and densities, ρ_1 and ρ_2 are chosen so that the barotropic
and first baroclinic modes of the model are similar to those of
the ocean. The lowest layer is the bottom mixed layer.
Similar models (without the bottom mixed layer) have been
successful in reproducing the statistics of eddies in the ocean
(e.g. Owens and Bretherton, 1978). The mixed layer height is
assumed to be restricted by stratification and capped by a
strong density gradient. A constant eddy viscosity, A_v, is
used in the boundary layer equations so that an explicit
equation can be derived for the mixed layer height.

   The equations for the quasi-geostrophic flow in the two
upper layers are

$$\frac{D_1}{Dt} [\nabla^2 \psi_1 + \beta y + F_1 (\psi_2 - \psi_1)] = q \qquad (3.1)$$

$$\frac{D_2}{Dt} [\nabla^2 \psi_2 + \beta y - F_2 (\psi_2 - \psi_1) + rh] = 0 \qquad (3.2)$$

where ψ_1, ψ_2 are the streamfunctions in the upper and lower
layers, β the rate of change of the Coriolis parameter (taken
to be a constant) and D_i/Dt the total derivative

$[\partial/\partial t - (\partial\psi_i/\partial y)\partial/\partial x + (\partial\psi_i/\partial x)\partial/\partial y]$. All length and velocity scales have been nondimensionalised with typical scales L and U, respectively, except the mixed layer height which is nondimensionalised with the Ekman depth $(2Av/f)^{\frac{1}{2}}$. The terms $F_1 = f^2 L^2/g' H_1$ and $F_2 = f^2 L^2/g' H_2$ where g' is the reduced gravity $(\rho_2 - \rho_1) g/\rho_1$ and $r = (2Avf)^{\frac{1}{2}} L/UH_2$. Equations (3.1) and (3.2) express the fact that the rate of change of relative vorticity in each layer is given by the rate of change of planetary vorticity plus the stretching of vorticity by the motions of the other interior layer. An additional vortex stretching in the second layer is due to the changing height of the mixed layer. A forcing, q, is introduced into the first layer which is such as to give a constant rate of energy input in a narrow wavenumber band.

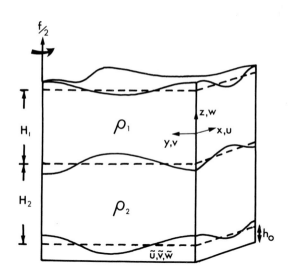

Fig. 3 Sketch of the regions of the flow of the model of
        Richards (1984).

To the same order of approximation the boundary layer equations for the mixed layer reduce to the steady state form of equations (2.1) and (2.2). The rate of change of the mixed height is obtained by considering the ageostrophic flux. Assuming a constant eddy viscosity, Av, the mixed layer height, h, is found to be given by

$$\frac{\partial h}{\partial t} + [-(1-G_1)\frac{\partial \psi_2}{\partial y} - G_2\frac{\partial \psi_2}{\partial x}]\frac{\partial h}{\partial x} + [(1-G_1)\frac{\partial \psi_2}{\partial x} - G_2\frac{\partial \psi_2}{\partial y}\frac{\partial h}{\partial y}]$$

$$= \tfrac{1}{2}b\nabla^2\psi_2 + u_e \qquad\qquad (3.3)$$

where

$$b = \frac{\sinh 2h - \sin 2h}{\cosh 2h + \cos 2h}$$

$$G_1 = \frac{2(1 + \cos 2h\cosh 2h)}{(\cosh 2h + \cos 2h)^2}$$

$$G_2 = \frac{2 \sinh 2h \sin 2h}{(\cosh 2h + \cos 2h)^2}$$

For a deep mixed layer (deep compared to the Ekman layer depth) the functions $b \to 1$, G_1 and $G_2 \to 0$. Substituting (3.3) into (3.2) gives a simple linear frictional term. For mixed layer heights of $O(1)$, however, G_1, and G_2 are non-zero and the advection velocity in the mixed layer becomes different from that of the flow above. This shear in advection velocities can significantly affect the dynamics of the system and can induce an instability (Richards, 1982b) or generate strong zonal flows (Richards, 1984).

When the thickness of the mixed layer is reduced by the action of eddies the top of the mixed layer is brought closer to the bottom. The turbulent energy at the interface between the mixed layer and interior increases and fluid is entrained into the mixed layer from above (see Fig. 4). In the model the mixed layer is allowed to mix up to a given equilibrium height on a timescale of 3 days. To counteract this, fluid is lost from the mixed layer either through separation of the layer in a manner described by Armi and D'Asaro (1980) or by ejection of fluid at fronts. Including these effects in the model is difficult due to the present lack of understanding of such events. As a compromise a constant detrainment rate is prescribed over the whole region which depends upon the timescale for matter to enter mix and leave the layer ('exchange time'). Measurements of heat fluxes suggest that the exchange time is between 4 and 600 days (Armi and D'Asaro, 1980). From their measurements of detatched mixed layers Armi and D'Asaro estimate the exchange time to be roughly 100 days.

The net rate of increase in height due to entrainment/
detrainment processes is modelled by the term u_e in (3.3).

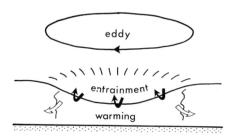

Fig. 4   Sketch of the interaction of an anticyclonic eddy with
         the bottom mixed layer. The thinning of the mixed
         layer allows entrainment of the interior fluid into the
         layer and a subsequent warming of the layer.

   Equations (3.1) to (3.3) are solved numerically (see
Richards (1984) for details). Typical stream function maps for
the first and second layers and the mixed layer height are
shown in Fig. 5 at a time after the model has reached a
statistically steady state. The eddies in the second layer
have an average speed of 4 cm s^{-1} and have a length scale of
50 km. The mixed layer height has small scale intense
features with regions of large gradient. The height of the
mixed layer varies between 0 and 100 m and has a horizontal
length scale of 35 m.

   Entrainment of the warmer fluid above the bottom mixed
layer into the mixed layer causes an increase in the mixed
layer temperature. Richards (1984) has investigated the
evolution of the temperature field of the mixed layer. The
results show that areas of warmer water which are surrounded
by regions of strong temperature gradients are very persistent,
lasting well over 100 days. At a single location the height of
the mixed layer is found to vary over a period of approximately
20 days with sharp increases in the temperature occurring.
Episodes of a similar structure and duration are seen in the
temperature records of Armi and D'Asaro (1980). It is possible
that the fronts observed by Thorpe (1983) could have also been
formed by the action of mesoscale eddies. The temperature
difference of 4 m$^{\circ}$C observed at a typical front requires a
vertical downward displacement of approximately 50 m.

   The observations of Armi and D'Asaro (1980) showed the
existence of interior layers uniform in temperature above the
bottom mixed layer. There is evidence to suggest that these
interior layers are formed by the detachment of the bottom

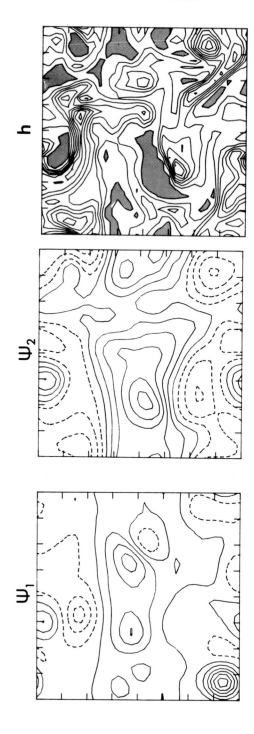

Fig. 5   Typical stream function maps for the flow in the upper two layers ψ_1, ψ_2 and the mixed layer height, h (after Richards, 1984).

mixed layer and may be the principal mechanism for fluid to escape from the bottom layer. The warm patches of bottom mixed layer produced by the action of eddies in the model of Richards (1984) are likely places of bottom layer detachment. These patches are lighter than the surrounding mixed layer and may be lifted off the bottom by buoyancy forces. The residence time for fluid to remain within the bottom layer therefore depends on the length of time it takes a fluid particle to come into contact with a warm patch. It is postulated that a fluid particle entering a warm area of the bottom mixed layer escapes from the layer.

Using the flow field within the mixed layer given by the model of Richards (1984) a number of particles (289) placed in the bottom mixed layer were tracked. The particles were assumed to be advected with the vertically averaged velocity within the mixed layer. The time was noted when each particle first entered an area of possible detatchment. Particles placed initially within a warm patch were immediately removed. The average residence time of the particles placed initially outside a detachment area was found to be approximately 800 days. As it is uncertain whether all warm patches will detach from the bottom this time should be treated as a lower bound on the residence time. The residence time for fluid in the regions where mixed layer is deeper and colder is therefore much greater than the average residence time for all particles of 100 days.

4. CONCLUSION

A number of processes controlling the height and horizontal variation of the bottom mixed layer of the ocean have been reported. The model described in Section 3 provides estimates of the residence time of fluid within the bottom mixed layer. To have confidence in the model results they have to be carefully tested with observations. The statistics of the eddies in the model can be matched to those observed. Validation of the model predictions of the formation of new mixed layers which may detach from the bottom is difficult, however, due to the fact that such structures will be advected with the flow.

So far, topography has not been included in models of the benthic boundary layer. Not only can topography affect the mesoscale eddy field but separation of the flow may take place leading to a much deeper mixed layer. The inclusion of topography in a model without flow separation is straightforward. Flow separation is much more difficult and will require a combination of numerical, laboratory and field experiments to examine the implications for mixing in the deep ocean.

ACKNOWLEDGMENT

This work has been carried out under contract for the
Department of the Environment, as part of its radioactive
waste management research programme. The results will be used
on the formulation of Government policy, but at this stage
they do not necessarily represent Government policy.

REFERENCES

Armi, L., and D'Asaro, E., (1980) Flow of the benthic ocean.
   J. Geophys. Res., **85**, 469-484.

Armi, L., and Millard, R.C., (1976) The bottom boundary layer
   of the deep ocean. *J. Geophys. Res.*, **81**, 4983-4990.

Bird, A.A., and Weatherly, G.L., (1982) A study of the bottom
   boundary layer over the eastward scarp of the Bermuda Rise.
   J. Geophys. Res., **87**, 7941-7954.

D'Asaro, E., (1982) Velocity structure of the benthic ocean.
   J. Phys. Oceanogr., **12**, 313-322.

Elliott, A.J. (1984) Measurements of the turbulence in an
   abyssal boundary layer. *J. Phys. Oceanogr.*, **14**, 1779-1786.

Hayes, S.P., (1979) Benthic current observations at Domes
   sites A, B and C in the tropical Pacific Ocean. In Marine
   geology and oceanography of the Pacific manganese nodule
   basin. Eds: J.L. Bishott and D.Z. Piper, Plenum Pub. Corp.
   83-112.

Holland, W.R., (1978) The role of mesoscale eddies in the
   general circulation of the ocean - numerical experiments
   using a wind-driven quasi-geostrophic model. *J. Phys.
   Oceanogr.*, **8**, 363-392.

Mellor, G.L., and Yamada, T., (1974) A hierarchy of turbulence
   closure models for planetary boundary layers. *J. Atmos.
   Sci.*, **31**, 1791-1806.

Owens, W.B., and Bretherton, F.P., (1978) A numerical study of
   mid-ocean mesoscale eddies. Deep-Sea Res. **25**, 1-14.

Richards, K.J., (1982a) Modelling the benthic boundary layer.
   J. Phys. Oceanogr., **12**, 428-439.

Richards, K.J., (1982b) The effect of a bottom boundary layer
   on the stability of a baroclinic zonal current. *J. Phys.
   Oceanogr.*, **12**, 1493-1505.

Richards, K.J., (1984)   The interaction between the bottom
    mixed layer and mesoscale motions of the ocean: a numerical
    study.   *J. Phys. Oceanogr.,* **14**, 754-768.

Saunders, P.M., (1983)   Benthic observations on the Madeira
    abyssal plain: currents and dispersion.   *J. Phys. Oceanogr.*
    13, 1416-1429.

Thorpe, S.A., (1983)   Benthic observations on the Madeira
    abyssal plain: fronts.   *J. Phys. Oceanogr.,* **13**, 1430-1440.

Weatherly, G.L., Blumsack, S.L., and Bird, A.A., (1980)   On the
    effect of diurnal tidal currents in determining the
    thickness of the turbulent Ekman bottom boundary layer.
    J. Phys. Oceanogr., **10**, 297-300.

Weatherly, G.L., and Martin, P.J., (1978)   On the structure and
    dynamics of the oceanic bottom boundary layer.   *J. Phys.
    Oceanogr.,* **8**, 557-570.

Wimbush, M., and Munk, W., (1970)   The benthic boundary layer.
    The Sea, *Vol. 4, Pt. 1, Wiley,* 731-758.

Wyngaard, J.C., (1975)   Modelling the planetary boundary layer -
    extension to the stable case.   *Boundary-Layer Meteor.,* **9**,
    441-460.

TURBULENCE AND WAVES IN STABLE LAYERS

S.J. Caughey

(Meteorological Office, Belfast)

ABSTRACT

Wave and turbulence subranges in temperature and velocity
spectra from the stable boundary layer (SBL) are identified
and discussed. The non-dimensional behaviour of the turbulence
subrange is examined as a function of height through the SBL
depth and the dependence on stability considered. The remark-
able variety of SBL turbulence structure is illustrated using
data from acoustic sounding.

REFERENCE

Caughey, S.J. (1982) Observed characteristics of the
    atmospheric boundary layer. Atmospheric turbulence and
    air pollution modelling. Ed. F.T.M. Nieuwstadt and H. van
    Dop, Reidel.

CALCULATION OF THE DEVELOPMENT OF THREE-DIMENSIONAL WAKE FLOWS IN A STABLY STRATIFIED ENVIRONMENT

J. McGuirk, A. Ghobadian, A.J.H. Goddard and A.D. Gosman
(Imperial College of Science and Technology)

ABSTRACT

The presentation described the preliminary calculations of the development of the wakes created by axi-symmetric bodies moving in an environment possessing a stable density stratification. Experiments have shown that initially round wakes undergo a gradual distortion in their shape as they evolve with downstream distance. The restoring action of the buoyancy force inhibits the vertical growth of the wake and enhances its horizontal spreading. At some downstream distance the wake reaches a maximum vertical size and the phenomenon of "wake collapse" occurs, as the mixed fluid seeks out, under the action of gravity, the level at which its density matches that of the environment. A mathematical model was described which allows the prediction of this phenomenon. The model is based on finite-difference solution of the 3D parabolic form of the momentum equations and uses an eddy viscosity based two equation turbulence model. Results will be described for both momentum-less and momentum deficit wakes in an ambient stable density stratification characterised via the Brunt-Vaisala period defined at

$$T = 2\pi/\{- \frac{g}{\rho} \frac{d\rho}{dz}\}^{\frac{1}{2}}$$

values of T between 50 and 600 seconds have been considered so far.

Points of interest which were discussed and which arise from these calculations include:-

(i) A demonstration that the experimentally observed values of the maximum wake height and the downstream distance at which it occurs are predicted well by the model used at a reasonable computational cost.

(ii)   The variation of these parameters with the densimetric
       Froude number F_D defined as

$$F_D = \frac{U_oT}{D_o} \qquad (U_o, D_o = \text{body speed, diameter})$$

       is also simulated well, but improvements to the
       turbulence model are believed necessary for very strong
       buoyancy influence ($F_D < 50$).

(iii) The modelling of the buoyancy effects in the turbulence
       model are shown to be important in the prediction of the
       wake collapse phenomenon. Calculations omitting these
       effects imply a much slower collapse.

(iv)   More experimental data in the form of velocity and
       density profiles are required to examine the performance
       of the model in more detail.

BIBLIOGRAPHY

1.   Gosman, A.D., Goddard, A.J.H., Ghobadian, A., and Nixon, W.,
     "Numerical simulation of coastal internal boundary layer
     development and comparison with simple models". Proc. 14
     Int. Technical Meeting in Air Pollution Modelling and its
     Applications, Copenhagen, Denmark, 1983.

2.   Benodekar, R.W., Goddard, A.J.H., and Gosman, A.D.,
     "Predicting lift-off of major self-heating releases under
     the influence of a building" Proc. CEC Seminar "Results of
     Indirect Action Research Programme" Safety Waters Reactors
     (1979-83), Brussels, October 1984.

3.   McGuirk, J.J. and Papadimitriou "Buoyant surface layers
     under fully entraining and internal hydraulic jump
     conditions" To be presented 5th Int. Turb. Shear Flows
     Symposium, Ithaca, August 1985.

3. DISPERSION IN STRATIFIED FLOW IN THE ENVIRONMENT

TOWARDS A BOX MODEL OF ALL STAGES OF HEAVY GAS CLOUD DISPERSION

P.C. Chatwin

(University of Liverpool)

ABSTRACT

The paper describes the philosophy of box models, widely used in predicting the potential hazards associated with the accidental release of heavy gas clouds. Attention is then focussed on the development of a box model describing all phases of dispersion, from release to the final stage of passive behaviour. Particular emphasis is placed on

(a) the effects of ambient wind shear,

(b) obtaining a model without unnatural abrupt transition between the different phases of dispersion,

(c) retaining the simplicity which is an important practical advantage of box models compared with other methods of prediction.

1. INTRODUCTION

1.1 Brief historical remarks

The work described in this paper was performed for the Health and Safety Executive (HSE) which, since 1976, has led a research programme to clarify the behaviour of massive releases of toxic or flammable gases. Discussion of this programme is given by McQuaid (1982), and recent summaries of the worldwide position are given, for example, in Havens (1982) and Raj (1982).

The HSE programme includes several series of experiments and, simultaneously, the investigation and development of predictive models. Among the series of experiments are 42 trials of $40m^3$ releases performed on Porton Down in 1976-1978 (Picknett 1978),

several series of wind tunnel simulations performed at Warren
Spring Laboratory during the period 1976-1982 (Hall, Hollis and
Ishaq 1982) and 10-15 trials of 2000m^3 releases performed on
Thorney Island in 1982-83. This last series of trials is
sponsored by organizations from many different countries, and
is still in progress at the time of writing (May 1983).

One main purpose of this experimental programme, and of
others being conducted elsewhere, is to provide a base of
reliable data for use in the validation of predictive models.

1.2   The different philosophies of models of heavy gas
dispersion

It is, at the outset, important to recognize that the
purpose of a model must be decided before it can be sensibly
developed. For example, since the dispersion of a heavy gas
cloud is a complicated matter, involving several different
interacting processes (turbulence in the ambient atmosphere,
turbulence generated by buoyancy forces, mixing across density
interfaces, etc.) it would undoubtedly be valuable to develop
a model giving a complete and accurate scientific description
of the dispersion. While such a model might ultimately have
some practical spin-off, it would not be likely to satisfy
fully the needs of many users of predictive models. First of
all, the development of this model might take a large number of
years so that other models of hazard assessment would be needed
meanwhile and, secondly, it is highly probable that, even when
developed, this full scientific model would be far too complicated for routine use in planning and siting of installations
and in emergency response measures.*

On the other hand, the work described in this paper is about
models with an immediate practical purpose, namely the prediction to acceptable accuracy, and with due regard to practical
convenience, of the distance from the site of the accidental
release at which the associated hazard ceases to exist. Since
this distance is a random variable it has been argued (Chatwin
1981a, 1982) that practical models ought to be stochastic.
However most existing models are deterministic, i.e. they do

---

*I tried to stress the importance of the points made in this
paragraph, particularly its last sentence, in response to some
questioners after my talk at the conference. Of course, I
agree with them that box models do not accurately describe the
structure of the cloud immediately after release, but do they
understand, conversely, that this criticism, valid from a
scientific point of view, does not relate to the urgent
practical needs of model users which box models do try to meet?

not take account of statistical fluctuations, and it is such
models that are of concern here. Broadly speaking, and with
admittedly some oversimplification, there are two groups of
such models.

The members of one group of models, of which MARIAH
(Deygon-Ra Inc.) and ZEPHYR (Energy Resources Co., Inc.) are
two examples, are based on numerical solutions of partial
differential equations governing the evolution in space and
time of quantities such as the ensemble mean velocity, and concentration, of the gas. These equations are approximate since,
among other things, they invoke eddy viscosities and diffusivi-
ties and, in addition, these transfer coefficients are related
to quantities like stratification on the basis of rather inadequate experimental data. It is unfortunate that thorough
evaluation of these models is not possible because their
detailed structures have not been made freely available.
However, based on statements in the open literature, e.g.
Woodward, Havens, McBride and Taft (1982) it appears that these
models are flexible in that, for example, they can incorporate
factors like non-uniform terrain, and are now, increasingly,
becoming practically cost effective. However their results are
not only subject to uncertainties, but also give far more
information than is often required for hazard assessment.
Furthermore it will not always be easy to have these programs
rapidly available on the site of an accident. There are
differing views on the future value of these models; it is
interesting to compare the views of Drake and Taft on p. 124
and pps. 139-140 respectively of Havens (1982).

However this interesting and important point is not pursued
further here since this paper is primarily concerned with
members of a second group of models, namely box models. Unlike
the models summarized above, box models do not attempt to
describe the detailed spatial variation of the gas velocity and
concentration. Rather, they assume that, to adequate accuracy
for practical purposes, the position of the gas cloud, and its
volume, can be predicted by supposing that the gas cloud has
the shape of a circular cylinder (the "box") within which the
gas has a uniform concentration. Box models consist of a set
of ordinary differential equations describing the evolution in
time of quantities like the position, radius and height of the
box, these equations being derived on the basis of simple
assumptions about the dominant physical processes. It is
important to understand - and this point relates to remarks
earlier in this section - that box models have the limited aim
of predicting only those properties of the gas cloud that are
of practical importance in hazard assessment and, therefore,
they ought to be judged only in terms of how well their predictions of these properties, and these properties alone, accord

with observations. The first box model was due to van Ulden
(1974) but the approach and philosophy are strongly reminiscent
of those adopted by Morton, Taylor and Turner (1956) in work on
turbulent convection which has now become classical. The same
ideas have also been applied in many subsequent studies, with
results consistent with experimental data.

1.3 The structure of existing box models

The basic situation considered in this section is the
instantaneous release (by sudden catastrophic loss of contain-
ment) of a finite volume of heavy gas of uniform density ρ_O
into an atmosphere with uniform air density ρ_a, where $\rho_O > \rho_a$.
Effects due to ambient atmospheric stratification will be
briefly considered later. While most existing box models
include thermodynamic equations which take account of initial
temperature (or initial phase) differences between the released
material and the ambient atmosphere, these will, for simplicity,
not be included in the following account, which therefore
relates to isothermal releases like the Porton and Thorney
Island trials. Neither will effects due to ground slope be
considered here, these having been treated by Britter (1982).

As explained above, existing box models assume that, for all
time t, the gas cloud has the shape of a circular cylinder with
a vertical axis perpendicular to the ground. Referring to
Fig. 1, the horizontal displacement of the cloud OO' at time t
since release will be denoted by $\bar{x}$, the radius and height of
the cylinder by r and h, and the gas density by ρ. A zero subscript, e.g. ρ_O, will be used to denote the initial value of
the variable. A second basic hypothesis of box models is that
the concentration C of the gas is spatially uniform within the
cylinder, C being measured in arbitrary units. By mass
conservation

$$\frac{C}{C_O} = \frac{v_O}{v} = \frac{r_O^2 h_O}{r^2 h} \ . \tag{1.3.1}$$

The basic aim of every box model is to predict how $\bar{x}$ and C (or
v) vary with t.

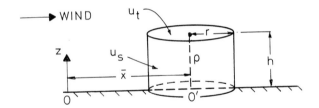

Fig. 1   Sketch for discussion of box models

Under isothermal conditions the total negative buoyancy of the cloud is conserved. Writing

$$g' = g\left\{ \frac{\rho - \rho_a}{\rho_a} \right\}$$ (1.3.2)

for modified gravity, it follows that

$$vg' = \pi r^2 hg' = \pi b_0,$$ (1.3.3)

where b_0 is a constant of dimensions $L^4 T^{-2}$. Fay and Ranck (1981) and Raj (1982) note that (1.3.3) will be a reasonable approximation in many circumstances where conditions are not isothermal.

All box models include an equation for the rate of spreading of the cloud about its axis, i.e. for dr/dt. The relationship used in most models is

$$\frac{dr}{dt} = \alpha (g' h)^{\frac{1}{2}},$$ (1.3.4)

where α is a constant of order unity. Equation (1.3.4) is based on the simple physical principle that the rate of spreading depends uniquely on the excess hydrostatic pressure within the cloud, and is consistent with observations from a broad range of experiments. Some models use g" instead of g' in (1.3.4), where $g'' = g(\rho - \rho_a)/\rho$. Though the difference is an important one in principle, discussed in detail by Fay (1982), it is unlikely to have great practical significance (van Ulden

1974). From (1.3.3) and (1.3.4) and the initial condition that
$r = r_O$ at $t = 0$, it follows (Picknett 1978) that

$$r^2 = r_O^2 \left\{ 1 + \frac{t}{t_O} \right\}$$

(1.3.5)

where

$$t_O = \frac{r_O^2}{2ab_O^{\frac{1}{2}}} ,$$

(1.3.6)

is a characteristic time scale, of order 0.3 - 0.5s for both
the Porton and Thorney Island trials.

As shown schematically in Fig. 1, box models assume that
mixing of the ambient atmosphere with the heavy gas takes place
over the surface of the box with entrainment velocities u_t and
u_s over the top and side surfaces of the cloud respectively.
Both u_t and u_s are assumed not to vary with position on the
appropriate surface. These assumptions lead to the following
entrainment equation for dv/dt:

$$\frac{dv}{dt} = \pi r^2 u_t + 2\pi rhu_s .$$

(1.3.7)

Existing models differ widely in the forms used for u_t and u_s,
reflecting the current lack of understanding of the fundamental
processes governing entrainment, particularly in the presence
of density interfaces. Models for which u_s is not identically
zero usually take u_s proportional (or simply related) to dr/dt,
given by (1.3.4). There is even less agreement about u_t, but
the later models all attempt to describe the tendency for the
stable density interface at the top of the cloud to inhibit the
vertical mixing (Turner 1979, Ch. 9). Thus u_t is taken to be
a decreasing function of a Richardson number. The function,
and the definition, of the Richardson number Ri, differ from
model to model. For immediate purposes it is sufficient to
note that in this paper Ri will be defined by

$$Ri = \frac{g' h}{u_*^2} , \qquad (1.3.8)$$

where u_* is the friction velocity, and also that, with this definition of Ri, the relationship

$$u_t \propto u_*(Ri)^{-1} \text{ for } Ri \gg 1 \qquad (1.3.9)$$

is consistent with that used in many models.

Once detailed formulae have been prescribed for u_t and u_s, (1.3.7) can be integrated to find v in terms of t making use of (1.3.5) and, of course, $v = \pi r^2 h$. Some existing models will be discussed in more detail later, but a comprehensive survey of all box models is given by Webber (1983).

Another ingredient of all box models is a prescription of how the cloud is advected as a whole by the mean wind $\bar{u}(z)$. In neutrally stable atmospheres

$$\bar{u}(z) = \frac{u_*}{\kappa} \ln\left\{ \frac{z}{z_0} \right\} , \qquad (1.3.10)$$

and box models either assume that $d\bar{x}/dt$ is constant, or that

$$\frac{d\bar{x}}{dt} = \bar{u}(\epsilon h) , \qquad (1.3.11)$$

where ϵ is a constant.

The discussion above assumes that the dispersion is dominated by gravity. However $g' \rightarrow 0$ as $t \rightarrow \infty$ so that the spreading and entrainment processes must in practice eventually be dominated by the motion of the ambient atmosphere. Accordingly, existing models either include a criterion giving abrupt transition into passive atmospheric dispersion described by a Gaussian model (Pasquill 1974), or they evolve gradually into box models of passive dispersion. The first option is unphysical and therefore unsatisfactory; the second option is therefore recommended and its consideration occupies much of the remainder of this paper.

2. A BOX MODEL OF ALL PHASES OF HEAVY GAS CLOUD DISPERSION

2.1 A box model of passive dispersion

   Those existing models which include gradual transition from
the gravity phase into the passive phase (e.g. Eidsvik 1980,
Fay and Ranck 1981) do so only by modifying the formula for u_t
in the entrainment equation (1.3.7), and not the formula for
u_s, or the spreading equation (1.3.4). Since, however, side
entrainment and spreading also occur throughout the dispersion
process, there seems little justification for singling out top
entrainment for special treatment. Therefore, the first step
taken in this section is to set up a box model of the passive
dispersion phase*, i.e. dispersion of the gas cloud when it has
effectively ambient density. This box model will then be used
later in setting up a consolidated box model describing all
phases of the dispersion.

   When the atmosphere is neutrally stable, the mean wind is
given by the logarithmic law (1.3.10), and the intensity of the
mixing at height z depends only on u_* (and z). With the
notation of section 1, it follows on dimensional grounds that
(1.3.4) - the spreading equation - must be replaced for passive
dispersion by

$$\frac{dr}{dt} = \alpha_1 u_*,\tag{2.1.1}$$

where α_1 is a constant, and that (1.3.5) is replaced by

$$r = r_0 \left[1 + \frac{\alpha_1 u_* t}{r_0} \right].\tag{2.1.2}$$

Similarly, the entrainment velocities u_t and u_s must both be
proportional to u_*, so that (1.3.7) becomes

$$\frac{dv}{dt} = \pi r^2 \beta_1 u_* + 2\pi r h \gamma_1 u_*,\tag{2.1.3}$$

---

* Superficially this box model may appear less sophisticated
than Gaussian models. However, it could be argued, and with
much force, that the models are in fact equivalent, since both
prescribe, rather than determine, the spatial variation of the
gas concentration.

with β_1 and γ_1 being further constants.   Using (2.1.1) and $v = \pi r^2 h$, it follows that

$$v = \left(\frac{\pi \beta_1}{3\alpha_1 - 2\gamma_1}\right) r^3 + \left\{v_0 - \left(\frac{\pi \beta_1}{3\alpha_1 - 2\gamma_1}\right) r_0^3\right\} \left(\frac{r}{r_0}\right)^{2\gamma_1/\alpha_1} \quad (2.1.4)$$

It is expected that the constants α_1, β_1 and γ_1 will all be of order unity, in which case $(2\gamma_1/\alpha_1) < 3$. For large times therefore, i.e. for

$$t \gg \frac{r_0}{u_*} , \quad (2.1.5)$$

using (2.1.2), the result in (2.1.4) can be approximated by

$$v \approx \left(\frac{\pi \beta_1}{3\alpha_1 - 2\gamma_1}\right) r^3 \approx \left(\frac{\pi \beta_1 \alpha_1^3}{3\alpha_1 - 2\gamma_1}\right) (u_* t)^3 . \quad (2.1.6)$$

It is interesting to note that for the Porton trials the time scale r_0/u_* in (2.1.5) was typically of order 10-20s compared with 0.3-0.5s for the gravity spreading time scale t_0 in (1.3.6).

The results above, derived via the box model format, are of course consistent with Lagrangian similarity (Batchelor 1964). Thus, while the approach above does not involve details of the variation of the mean wind with height, its results have the same dependence on $u_* t$, when (2.1.5) is satisfied, as other approaches (with different purposes) which do include such details (e.g. Chatwin 1968, Hunt and Weber 1979). It follows that, as far as the values of v and C at large times are concerned, it would be a waste of time to attempt to "improve" the box model presented above to take explicit account of the detailed mean wind profile since, to be consistent with basic physics (i.e. Lagrangian similarity), the end result would inevitably be (2.1.6).

2.2   The effects of ambient wind shear on side entrainment for
      Ri >> 1

It was noted earlier that existing box models differ greatly
in their prescriptions of the entrainment velocities u_t and u_s
introduced in equation (1.3.7) for dv/dt. The following treatment of modelling side entrainment particularly emphasises the
effects of ambient wind shear, which no existing model takes
explicitly into account.

The term involving u_s in (1.3.7) presents the net effect of
two distinct mechanisms causing horizontal mixing of the heavy
gas and the ambient atmosphere. One such mechanism is
straightforward horizontal turbulent diffusion represented by
the terms

$$- \frac{\partial}{\partial x} \overline{(u' c')} - \frac{\partial}{\partial y} \overline{(v' c')}$$

in the standard equation, where u', v' and c' are the fluctuations about the ensemble means of the x and y components of
velocity, and the concentration respectively. This is
indicated schematically in Fig. 2(i). The second mechanism
causing spreading, namely shear dispersion, is indicated in
Fig. 2(ii). As the result of the mean shear, the cloud is
itself sheared in the x direction. Acting by itself, this
would lead to a rate of spreading of the cloud in the x direction, proportional to $\bar{u}(h_0)$. However vertical gradients of C
are increased (indeed created for the particular initial shape
considered here) by this shearing of the cloud. Therefore,
vertical turbulent diffusion, represented by the term

$$- \frac{\partial}{\partial z} \overline{(w' c')}$$

in the standard equation for C, is greatly enhanced. As a
result the wind shear is immediately less effective in
spreading the cloud horizontally than it would be in the
absence of vertical diffusion. Nevertheless, vertical diffusion does not prevent all spreading due to the mean wind, and
it is the combined effect of the two processes shown in Fig.
2(ii) which is shear dispersion, the second of the two
mechanisms referred to above.

For the dispersion of a passive cloud in an unbounded
neutrally stable atmosphere, the rate of spread, i.e. u_s in

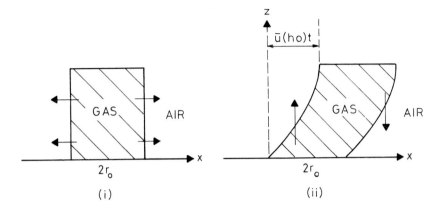

Fig. 2 Mechanisms causing horizontal entrainment
        (i)   direct horizontal dispersion;
        (ii) direct effects of wind shear are reduced
              by enhanced vertical dispersion-shear
              dispersion.

(1.3.7), for each of the mechanisms is proportional to u_* (Batchelor 1964, Chatwin 1968, Yaglom 1976), but such simplicity cannot be expected to apply a priori for heavy clouds. As far as direct horizontal diffusion is concerned, there will be contributions due to turbulence in the ambient atmosphere and to turbulence generated by the negative buoyancy of the gas cloud. On dimensional grounds these can be expected to make contributions to u_s proportional to u_* and $(g'h)^{\frac{1}{2}}$ respectively. More generally, considering the inevitable interactions between both sources of turbulence, a reasonable proposal for the direct horizontal diffusion contribution to u_s is

$$u_s = (g'h)^{\frac{1}{2}} f(Ri), \qquad (2.2.1)$$

where Ri is defined in (1.3.8). As the dispersion progresses, Ri must inevitably decrease, and (2.2.1) must reduce to $u_s \propto u_*$, consistent with the discussion in §2.1. Thus

$$Ri \ll 1 \implies f(Ri) \propto (Ri)^{-\frac{1}{2}}. \qquad (2.2.2)$$

However, for heavy gas releases of interest, initial values of
Ri are inevitably large (e.g. $Ri_0 \sim 10^2 - 10^3$ for both the
Porton and Thorney Island trials) and the value of u_s must be
dominated by gravity effects so that

$$Ri \gg 1 \Rightarrow f(Ri) \approx constant. \qquad (2.2.3)$$

In turning now to the effect of shear dispersion, it is
important to stress first that wind shear, acting by itself,
does not change the volume of the region occupied by heavy gas,
but only the shape of this region. As anticipated above,
vertical mixing must be included in the assessment of wind
shear on dv/dt. As indicated schematically in Fig. 3, the
upstream edge of the cloud tends to become stably stratified,
with consequent inhibition of vertical mixing, while, conversely, vertical mixing is likely to be vigorous on the downstream edge. Consistent with the approach of box models, it
seems natural to represent the mixing across the sides of the
sheared cloud by an entrainment velocity $\underset{\sim}{u}_s$ directed along the
local normal to the cloud surface. Let $\underset{\sim}{u}_s$ have horizontal and
vertical components u_{sh} and u_{sv} respectively, so that the rate
of entrainment across a surface element of area δA is

$$|\underset{\sim}{u}_s| \delta A = (u_{sh}\cos\theta + u_{sv}\sin\theta) \delta A, \qquad (2.2.4)$$

where θ is the angle of inclination of the local normal to the
horizontal. Following the assumptions made in many box models
about u_t - the top entrainment velocity - as expressed in
(1.3.9), it is reasonable to suppose that on the upstream
stably stratified edge of the cloud, u_{sv} is given by

$$upstream: u_{sv} = \gamma' \left\{\frac{u_*}{Ri}\right\}, \qquad (2.2.5)$$

where γ' is a constant. On the downstream edge, however, u_{sv}
cannot depend significantly on u_* because the mixing there will
be vigorous, and dominated by the unstable stratification. The
speed of free fall from the top of the cloud in the presence of
this stratification is of order $(g'h)^{\frac{1}{2}}$. Although this is likely
to be an overestimate for u_{sv} near the bottom of the cloud, the
order of magnitude should be given by:

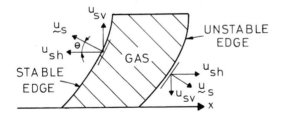

Fig. 3   Shear dispersion for heavy gas clouds

$$\text{downstream: } u_{sv} = \gamma''(g' h)^{\frac{1}{2}}. \tag{2.2.6}$$

Both (2.2.5) and (2.2.6) are, of course, estimates for Ri >> 1 and, for Ri << 1, they will both be replaced by $u_{sv} \propto u_*$ as in §2.1. For Ri >> 1, u_{sh} should be given, both upstream and downstream, by (2.2.1) and (2.2.3), so that $u_{sh} = \gamma'''(g' h)^{\frac{1}{2}}$. Thus (2.2.4) gives the rate of entrainment across δA when Ri >> 1 as

$$\text{upstream: } (g' h)^{\frac{1}{2}}\left\{\gamma''' \cos\theta + \frac{\gamma' \sin\theta}{(Ri)^{3/2}}\right\}\delta A; \tag{2.2.7}$$

$$\text{downstream: } (g' h)^{\frac{1}{2}}\{\gamma''' \cos\theta + \gamma'' \sin\theta\}\delta A.$$

While (2.2.7) does not consider other parts of the side of the cloud except the upstream and downstream edges, and while γ''' may be different in the two parts of (2.2.7), the following two conclusions seem legitimate:

(i)   the contribution of shear dispersion to the rate of entrainment over a small surface element is, for Ri >> 1 and, also, (as shown in §2.1) for Ri << 1, at most of the same order as that due to direct horizontal diffusion;

(ii)   for Ri >> 1, the rate of entrainment over a small surface element of area δA is of order $(g' h)^{\frac{1}{2}}\delta A$.

These conclusions refer to the local rate of entrainment
only. However there are other effects of shear on the gas
cloud. For example, the asymmetry of the vertical mixing
caused by shear, illustrated by the difference between (2.2.5)
and (2.2.6), will cause the cloud to "fill in" below its downstream edge and develop an asymmetric elevation. This tendency
was evident in the early stages of many of the Porton trials
(Picknett 1978) and in the laboratory simulation of some of
these trials (see especially Figs. 22, 31, 45 and 58 of Hall,
Hollis and Ishaq (1982)). Also, although shear does not itself
(i.e. without vertical mixing) change the volume of the cloud,
it does increase   the surface area of the side of the cloud
thereby tending to increase the overall rate of entrainment.
Both of these additional effects of wind shear will be greatest
near the beginning of the dispersion period before the height
to width ratio has become very small, and while Ri is still
large. Since experimental evidence (van Ulden 1974, Picknett
1978) shows that the overall rate of entrainment is most
intense immediately after release, and since this phenomenon is
not adequately modelled by existing box models, it therefore
seems reasonable to attempt an ad hoc amendment to model the
effects of developing asymmetry and increased side surface
area.

Fig. 4 shows schematically the geometrical ideas of this
amendment. The vertical cylinder of existing box models is
replaced by a cylinder of identical circular cross-section and
vertical height, but with an axis inclined at an angle $\phi = \phi(t)$
to the vertical. To adequate practical approximation, the side
surface area of the tilted cloud can be taken as $2\pi rh\sec\phi$
(Chatwin 1983), and the entrainment equation (1.3.7) is then
amended to

$$\frac{dv}{dt} = \pi r^2 u_t + 2\pi rh\sec\phi u_s.$$ (2.2.8)

It follows that box models with $u_s \equiv 0$ cannot be simply amended
in this way to take account of the increased surface area.
Indeed Fay and Ranck (1981) note that their model with $u_s \equiv 0$
may therefore "not be accurate at the earliest times" when the
cloud's height to width ratio is largest (and therefore when
the effect being considered here is potentially most important).

It remains to develop an equation for $\phi(t)$. This will
clearly depend on the difference in velocity between the top
and bottom of the cloud and, in particular, therefore on u_* and
z_0. Calculations presented in Chatwin (1983) and

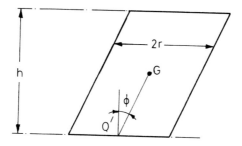

Fig. 4 The tilted cylinder of the amended box model

supported by some isolated experimental results from Hall, Hollis and Ishaq (1982) suggest that an appropriate formula for times of interest is

$$\tan\phi = \delta\left\{\frac{u_*t}{h}\right\}\ln\left\{\frac{h}{ez_0}\right\},$$
(2.2.9)

where e is the base of natural logarithms and δ is a constant to be determined from experimental observations, whose value seems likely to be of order 5.

Equations (1.3.5), (2.2.8) and (2.2.9), together with $v = \pi r^2 h$ and prescriptions of u_t and u_s form a closed system. Table 1 summarizes, for three existing models, the differences between the basic model and the amended model (i.e. the model using the same prescriptions of u_t and u_s as the basic model, but with (2.2.8) and (2.2.9) replacing (1.3.7)) for values of t/t_0 from 0 to 8, where t_0 is the time scale characterizing initial spreading defined in (1.3.6). In these calculations δ in (2.2.9) was taken as 5.

Table 1 shows, as anticipated, that the amended models predict substantially greater entrainment in all cases than the basic models. However, although the behaviour is qualitatively as required, the results are not likely to be realistic for other than the early times considered in Table 1. In the first place, all observations show that the geometrical assumptions underlying the amendment rapidly become untenable; indeed the values of ϕ for the results shown in Table 1 when $t/t_0 = 8$ range from $63°$ (for the Picknett model on Porton trial 33) to $87°$

Table 1

Comparisons between results of some basic box models, and their results when amended by (2.2.8) and (2.2.9). The results for the amended models were obtained by R. Fitzpatrick on the HSE computing facilities. Details of the basic models are given in van Ulden (1974), Picknett (1978), and Fryer and Kaiser (1979) with modifications to the last in Fitzpatrick (1981). Note that for the small values of t/t_O and high values of Ri_O in this table, the predicted differences between the two trials are too small to show up for each of the basic models above. Nor do the recorded data of the Porton trials allow comparison between the above predictions and experimental observations.

		t/t_O	O	2	4	6	8
	van Ulden	Basic v/v_O	1	1.06	1.08	1.10	1.12
		Amended v/v_O	1	1.09	1.28	1.57	1.98
(A)	Picknett	Basic v/v_O	1	2.46	3.74	4.93	6.06
		Amended v/v_O	1	3.12	7.70	15.5	26.8
	Modified DENZ	Basic v/v_O	1	1.93	2.63	3.21	3.74
		Amended v/v_O	1	2.38	5.27	10.17	17.29

PORTON TRIAL 3: $t_O \approx 0.26s$, $Ri_O \approx 464$, $\dfrac{h_O}{ez_O} \approx 644$

		t/t_O	O	2	4	6	8
	van Ulden	Basic v/v_O	1	1.06	1.08	1.10	1.12
		Amended v/v_O	1	1.06	1.13	1.23	1.35
(B)	Picknett	Basic v/v_O	1	2.46	3.74	4.93	6.06
		Amended v/v_O	1	2.58	4.64	7.61	11.7
	Modified DENZ	Basic v/v_O	1	1.93	2.63	3.21	3.74
		Amended v/v_O	1	2.01	3.23	4.98	7.40

PORTON TRIAL 33: $t_O \approx 0.28s$, $Ri_O \approx 1830$, $\dfrac{h_O}{ez_O} \approx 129$

(for the van Ulden model on Porton trial 3). More importantly, as already indicated, it is desirable to consider how results like (2.2.8) and (2.2.9) might be themselves amended so that there is smooth transition into the passive phase box model of §2.1.

2.3 The advection of the cloud

One further reason for caution in attaching quantitative significance to results like those shown in Table 1 is that no account is taken in box models of the effect of the initial shape of the cloud. Justification for this neglect is the need for a practically viable model together with the fact that, from the point of view of hazard assessment (which is the purpose of box models), times of interest are invariably much larger than those considered in Table 1*. Another effect of initial conditions, namely the delay time t_1 taken for the gas cloud to attain an advection velocity simply related to the profile of the mean wind is relevant in this section. A simple order of magnitude argument, essentially that of Cox and Carpenter (1980), gives

$$t_1 \sim \frac{\pi}{2C_D} \left(\frac{r_0}{\bar{U}}\right) \frac{\rho_0}{\rho_a} , \qquad (2.3.1)$$

with $\bar{U}$ being the mean wind and C_D being a drag coefficient. With $C_D = \frac{1}{2}$, the above estimate of t_1 is 3s for Porton trial 3, and 4s for Porton trial 33. These are the two trials used in deriving Table 1. As explained above, it is sensible to neglect such short times in developing practically useful models of hazard assessment.

Existing box models differ in their prescriptions of $\bar{x}$ in Fig. 1 as a function of t. Some (e.g. Picknett 1978, Eidsvik 1980) take $d\bar{x}/dt$ to be a constant fraction of the mean wind speed $\bar{U}$. However it is difficult to know how the mean wind speed should be precisely defined, and such prescriptions do not allow for the large relative changes in cloud height that are observed to occur. From the point of view of basic physics a more satisfactory proposal is to prescribe $d\bar{x}/dt$ as a weighted average of the mean wind profile over the cloud height:

---

*It is however understood that further investigation of the importance of initial cloud shape is currently being undertaken on behalf of HSE.

$$\frac{d\bar{x}}{dt} = \frac{\left\{\int_0^h F(z)\bar{u}(z)\,dz\right\}}{\left\{\int_0^h F(z)\,dz\right\}}. \tag{2.3.2}$$

It is natural to choose $F(z)$ to be proportional to the vertical variation of the ensemble mean concentration. For passive releases from a point source this choice leads to

$$\frac{d\bar{x}}{dt} = \bar{u}(\epsilon' \bar{z}) \tag{2.3.3}$$

for large times after release (Batchelor 1964), where $\bar{z}$ is the height of the cloud centroid and ϵ' is a constant estimated to be about 0.56 (Chatwin 1968).

For heavy gas clouds it is more conventional to use results involving the cloud height h rather than $\bar{z}$ and, accordingly, to replace (2.3.3) by

$$\frac{d\bar{x}}{dt} = \bar{u}(\epsilon h), \tag{2.3.4}$$

where ϵ is another constant. This formula is used in several box models with various values of ϵ. For example, van Ulden (1974) takes $\epsilon = e^{-1} \approx 0.37$, DENZ (Fryer and Kaiser 1979) takes $\epsilon = 0.5$ and Fay and Ranck (1981) take $\epsilon = 0.4$. The choice made by van Ulden of $\epsilon = e^{-1}$ is that given by (2.3.2) with $\bar{u}(z)$ given by the logarithmic law (1.3.10) with $F(z) = 1$, the functional form consistent with the assumption common to all box models that the distribution of concentration is uniform within the cloud. The same value of ϵ is also given for a passive cloud of height h in the very early stages of release (Chatwin 1983). For large times (2.3.3) and (2.3.4) must be compatible. If it is assumed that $\bar{z} = \frac{1}{2} h$ (although (Chatwin and Sullivan 1981) the factor of $\frac{1}{2}$ is probably somewhat too high) then (2.3.3) and (2.3.4) require $\epsilon = \frac{1}{2}\epsilon' = 0.28$ for compatibility. In view of the weak (logarithmic) dependence of $\bar{u}(z)$ upon z the results are not very sensitive to the choice of ϵ in (2.3.4). For example

$$\left.\frac{d\bar{x}}{dt}\right|_{\epsilon=0.37} - \left.\frac{d\bar{x}}{dt}\right|_{\epsilon=0.28} = \frac{u_*}{\kappa} \ln\left(\frac{0.37}{0.28}\right) \approx 0.7u_*, \quad (2.3.5)$$

which is about 5% of the value of $d\bar{x}/dt$ with $\epsilon = 0.37$, $h = 2m$ and $z_0 = 0.002m$. A reasonable compromise that retains (2.3.4) is to take $\epsilon = (0.28 \times 0.37)^{\frac{1}{2}} \approx 0.32$, i.e. to prescribe

$$\frac{d\bar{x}}{dt} = \bar{u}(0.32h). \quad (2.3.6)$$

It should be noted that, unlike the proposals made in §2.1 and §2.2, equation (2.3.6) is intended to cover the whole dispersion period from release until the final stages of passive dispersion. If interest is confined to a limited range of times a different value from 0.32 might well be appropriate. Thus Fay and Ranck (1981) demonstrate satisfactory agreement with experiments when $t/t_0 \lesssim 400$ and $\epsilon = 0.4$ in (2.3.4).

2.4 The transition to passive behaviour

As noted in §1.3, many existing box models (e.g. van Ulden (1974), Picknett (1978), DENZ (Fryer and Kaiser 1979)) include a criterion (or criteria) giving abrupt transition between a simple model of the early gravity-dominated stage of dispersion and another simple model of the final passive stage of dispersion. The early models of van Ulden (1974) and Picknett (1978) postulated transition when the spreading rate dr/dt, given by (1.3.4), became a fixed multiple of that given by (2.1.1)*. This leads to transition occurring when

$$(g' h)^{\frac{1}{2}} = Ku_*, \quad (2.4.1)$$

where $K = 2$ for van Ulden (1974) and $K = 2.66$ for Picknett (1978). Later models determine when transition occurs in different, generally more complicated ways. Thus transition occurs in DENZ when either

---

*Actually Picknett (1978) used a different physical argument, based on the penetration of atmospheric turbulence fluctuations into the cloud, but the result is equivalent to (2.4.1), and hence to the physical argument leading to it.

$$\rho - \rho_a < 10^{-3} \text{kg m}^{-3},$$

or when both of two other criteria are satisfied, the most restrictive of which is that u_t must be greater than a longitudinal velocity scale proportional to u_*.

Even though they are simple, such transition criteria are unsatisfactory because the actual evolution from one stage to another is of course gradual (see, for example, the discussion in Chatwin (1981a, 1981b)). There is also some evidence that the results of box models are rather sensitive to the precise criterion chosen. However it has now been established by Eidsvik (1980) and Fay and Ranck (1981), at least to some extent, that both of these objections can be overcome by devising box models that remain practically viable but include automatic gradual transition from the gravity-dominated stage into the passive stage.

2.5 The basis of a box model for all stages of dispersion in neutrally buoyant atmospheres

Both of the models referred to at the end of the last section emphasize the desirability of gradual transition only for their prescriptions of the top entrainment velocity u_t. However it now seems appropriate, in anticipation of the forthcoming availability of the data from the Thorney Island trials, to propose a model with gradual transition which incorporates the earlier conclusions of this paper, the most important of which are:

(i)   the effects of wind shear on side entrainment velocities do not need to be taken separately into account since they can be described using the standard parametrization (§2.2);

(ii) it is possible to model the effect of wind shear in increasing the effective side surface area of the cloud over which entrainment takes place (§2.2);

(iii) in view of the weak dependence of the mean wind profile on height it is likely that a formula like (2.3.6) will prove adequate for all stages of the dispersion (§2.3).

As noted by Webber (1983), the physics of the dispersion process is at its most complicated during the transition period and modelling the process gradually requires implicit assumptions to be made about the physics. Since experimental data of sufficient quality and quantity are not yet available to validate (or invalidate) any such assumptions, the most sensible procedure at the moment is to keep the model as simple as possible. All available evidence suggests that the single most important parameter is the Richardson number Ri defined in (1.3.8) or, equivalently, a Froude number proportional to $(Ri)^{-\frac{1}{2}}$. Note, for example, that (2.4.1) is equivalent to $Ri = K^2$, and that the model of Fay and Ranck (1981) is gravity-dominated for $Ri \gg 0.2$ and predicts passive dispersion for $Ri \ll 0.2$. Therefore the viewpoint adopted in the sequel is to incorporate transition by means of empirical formulae involving Ri which give correct dependence on the model parameters for both high and low values of Ri.

With this philosophy, consider first the rate of spreading of the cloud dr/dt. It is required that (1.3.4) holds for $Ri \gg 1$ and that (2.1.1) holds for $Ri \ll 1$. One of the simplest formulae consistent with both of these is

$$\frac{dr}{dt} = \alpha (g' h)^{\frac{1}{2}} \left\{ 1 + \frac{\alpha_1}{\alpha} (Ri)^{-\frac{1}{2}} \right\} , \qquad (2.5.1)$$

and it happens that this equation can be integrated analytically to give

$$\delta_* \left\{ \frac{r}{r_0} - 1 \right\} - \ln \left\{ \frac{1 + \delta_* \left(\frac{r}{r_0} \right)}{1 + \delta_*} \right\} = \frac{1}{2} \delta_*^2 \left(\frac{t}{t_0} \right) , \qquad (2.5.2)$$

where t_0 is defined in (1.3.6) and

$$\delta_* = 2\alpha_1 \left\{ \frac{u_* t_0}{r_0} \right\} . \qquad (2.5.3)$$

For all the Porton trials and all the Thorney Island trials $\delta_* \ll 1$. For example, with $\alpha_1 = 1$, $\delta_* \approx 0.046$ for Porton trial 3 and $\delta_* \approx 0.024$ for Porton trial 33. It is straightforward to verify that (2.5.2) has the following consequences:

$$\delta_* \left\{ \frac{r}{r_0} \right\} \ll 1 \Rightarrow r^2 = r_0^2 \left\{ 1 + \frac{t}{t_0} \right\} , \qquad (2.5.4)$$

which is precisely (1.3.5), and

$$\delta_* \left\{ \frac{r}{r_0} \right\} >> 1 \Rightarrow r = r_0 \left\{ 1 + \frac{\alpha_1 u_* t}{r_0} \right\}, \qquad (2.5.5)$$

which is precisely (2.1.2). Furthermore the condition $\delta_* (r/r_0) << 1$ can be rearranged to give $(Ri)^{\frac{1}{2}} >> \alpha_1$ using (1.3.3), and the above results are therefore consistent with the physical arguments leading to (2.4.1). Further discussion of (2.5.2) is given in Chatwin (1983).

The next, and most difficult, process to consider is entrainment. There is no reason not to base the discussion on the conventional formulation (1.3.7), namely

$$\frac{dv}{dt} = \pi r^2 u_t + 2\pi r h u_s, \qquad (2.5.6)$$

where the entrainment velocities u_t and u_s are to be taken as functions of u_*, Ri and (in the case of u_s) ϕ. For the large values of Ri occurring in the early stages of dispersion, u_t is usually taken to be $\beta u_* (Ri)^{-1}$ as in (1.3.9)*, and for small values of Ri it must be consistent with (2.1.3), i.e. $u_t = \beta_1 u_*$. Formulae consistent with these limits have been given both by Eidsvik (1980) and Fay and Ranck (1981). Eidsvik's formula is

$$u_t = \frac{\beta \beta_1 u_*}{\beta + \beta_1 (Ri)}, \qquad (2.5.7)$$

and Fay and Ranck's formula is

$$u_t = \frac{\beta \beta_1 u_*}{\{\beta^2 + \beta_1^2 (Ri)^2\}^{\frac{1}{2}}}. \qquad (2.5.8)$$

---

* However (Chatwin 1983) there is a case (see e.g. Turner 1979, p. 298) for taking $u_t \propto u_* (Ri)^{-3/2}$ with consequent amendments to later formulae, e.g. (2.5.7).

Table 2

Summary of proposed box model in dimensionless form where $R = r/r_0$, $V = v/v_0$, $T = t/t_0$ and t_0 is given by (1.3.6) as a function of the initial conditions. Note also that $Ri = \epsilon_* R^{-2}$ and that the solution of the equation for R^2 is given in equation (2.5.2).

$$\frac{dR^2}{dT} = (1 + \delta_* R)$$

$$\frac{dV}{dT} = R^2 \left(\frac{u_* t_0}{h_0}\right) \frac{u_t}{u_*} + \frac{2V}{R} \left(\frac{u_* t_0}{r_0}\right) \frac{u_s}{u_*} \quad ,$$

with $\quad \dfrac{u_t}{u_*} = \dfrac{\beta_1 R^2}{R^2 + (\beta_1 \epsilon_*/\beta)} \quad$ and $\quad \dfrac{u_s}{u_*} = \dfrac{\gamma_1 R^2 + (\gamma \epsilon_*/\gamma_*)}{R^2 + (\epsilon_*^{\frac{1}{2}}/\gamma_*) R\cos\phi} \quad ,$

where $\quad \tan\phi = \delta \left(\dfrac{u_* t_0}{h_0}\right) \dfrac{R^2 T}{V} \ell n \left(\dfrac{V}{R^2} \dfrac{h_0}{ez_0}\right), \quad \delta_* = 2\alpha_1 \left(\dfrac{u_* t_0}{r_0}\right) \quad ,$

$$\epsilon_* = (\alpha_1/\alpha\delta_*)^2 .$$

$$\frac{d\bar{x}}{dT} = \frac{u_* t_0}{\kappa} \ell n \left(\frac{V}{R^2} \frac{\epsilon h_0}{z_0}\right) .$$

These formulae will have very similar effects on the solution of (2.5.6), and it is not likely that decisive experimental evidence in favour of one or the other will be available soon. For present purposes (2.5.7) will be adopted on the grounds that it is marginally simpler. Turning now to side entrainment, u_s will be equal to $\gamma(g'h)^{\frac{1}{2}}\sec\phi$ for large values of Ri, according to the work in §2.2 and to $\gamma_1 u_*$ for small values of Ri, according to the work in §2.1. The simplest of many formulae consistent with these remarks that has been found is

$$u_s = \left\{ \frac{\gamma(\text{Ri})+\gamma_1\gamma_*}{\cos\phi(\text{Ri})+\gamma_*(\text{Ri})^{\frac{1}{2}}} \right\} (g'h)^{\frac{1}{2}}, \qquad (2.5.9)$$

where γ_* is another constant, and ϕ will be given by (2.2.9). The role of γ_* is to give an effective but gradual transition criterion since, according to (2.5.9), side entrainment is gravity-dominated when $(\text{Ri})^{\frac{1}{2}} \gg \gamma_*\sec\phi$ and is as in passive dispersion when $(\text{Ri})^{\frac{1}{2}} \ll \gamma_*\sec\phi$. It is important to note that, unlike (2.2.8) and (2.2.9), equations (2.5.9) and (2.2.9) will not give unrealistic difficulties when used in (2.5.6) because of the large values of $\tan\phi$ that rapidly occur.

The final equation in the proposed box model is (2.3.6) giving the advection of the cloud as a whole. Table 2 summarizes the proposed box model as a whole in dimensionless form, and Table 2 makes some comments on the constants appearing in Table 2.

Of the constants in Table 3, all those in Greek characters are universal if the proposed model is to have practical use, in the sense that the same values of these constants should give acceptable results over a wide range of spill geometries, initial densities and ambient atmospheric conditions.

3.   THE USE OF EXPERIMENTAL DATA

3.1 Introduction

It has already been noted that the data available at present do   not allow a comparison in depth of the results of the proposed model with experiments, certainly not to the extent of permitting validation of the separate steps in the modelling, namely spreading, side and top entrainment and advection.   Since the situation regarding available

Table 3

The constants appearing in the box model of Table 2.

$\alpha, \beta, \gamma;\ \alpha_1, \beta_1, \gamma_1$	All likely to be of order 1. Experiments suggest that $\alpha = 1$ is adequate and Fay and Ranck (1981) suggest $\beta_1 \approx 2.5$ on the basis of experiments.
δ	Calculations (Chatwin 1983) suggest $\delta \approx 5$.
ϵ	Equation (2.3.6) gives $\epsilon \approx 0.32$ as reasonable.
$\dfrac{u_* t_O}{h_O},\ \dfrac{u_* t_O}{r_O}$	Both determined by release and normally very small.
h_O / z_O	Determined by release. Often of order 30.

experimental data will improve markedly with the release of results from the Thorney Island trials, it is appropriate to make some observations regarding the use of these data.

3.2 *Some comments on the validation of box models using experimental data*

It should be recalled first that, for a variety of reasons, box models cannot be expected to be accurate in the very early stages of dispersion. Thus no box model incorporates details of the initial cloud shape other than the initial aspect ratio (r_O / h_O) and all box models ignore the effect of the initial cloud inertia in calculating the rate of advection. As a consequence, it is unlikely that data taken in the very early stages of dispersion (e.g. for $T = t/t_O$ less than 10-20) can be used to test the adequacy of the modelling of the wind shear effect in increasing the side surface area of the cloud, this modelling being through the terms in ϕ in Table 2. Such testing can be carried out only indirectly.

Furthermore the earlier discussion suggests that the effects of wind shear on the dispersion can most usefully be considered by regarding them, not as separate from, but as modifications

to, the basic processes. With the exception of Fay and Ranck
(1981), proposers of recent box models have not apparently
tested their models, particularly the descriptions of these
basic processes, against all the available data. Accordingly
the most urgent question is still how well box models can
describe these processes which suggests that the Thorney Island
data should be used in conjunction with the other available
data to investigate this question first. It may, or may not,
then become apparent that, for practical purposes at any rate,
one or other of the existing models gives acceptable accuracy,
at least over the range of conditions covered in the
experiments.

Some points which may be pertinent in such an investigation
can be identified:

(i)   In view of the legacy of initial conditions upon the
      subsequent dispersion, it is sensible to compare the
      data, not only with the predicted values of r, v and $\bar{x}$,
      but also with the basic equations of the box models in
      which the rates of change of these quantities are pre-
      dicted. It is certainly possible for the data to be
      consistent with these basic equations, but also not to
      be consistent to an acceptable accuracy with the
      formulae obtained by integration and fitting the initial
      conditions $r = r_0$, $v = v_0$ and $\bar{x} = 0$ at $t = 0$. In many

      other fields such as longitudinal dispersion (Chatwin
      1970, 1971) it is necessary to allow for the legacy of
      the initial conditions by, for example, using a virtual
      time origin whose value is dependent upon the initial
      conditions, but not simply dependent.

(ii) Advection is dealt with separately from the other basic
      processes in box models, and it is also the process in
      which (obviously) wind shear must be important. Since
      the value of $\bar{x}$ can probably be estimated from experi-
      ments rather more reliably than either r or v it would
      seem (relatively) straightforward to test whether
      (2.3.4), viz. $d\bar{x}/dt = \bar{u}(\epsilon h)$, is an adequate representa-
      tion of the data and, if so, to estimate the value of
      the constant ϵ. However it needs noting that the maxi-
      mum values of $u_* t/h$ in most experiments are not high, so
      that rather little passive dispersion has occurred. The
      consequence of this is that in assessing hazards such as
      toxicity, when very small concentrations can be
      extremely important, it may be necessary, as explained
      following (2.3.6) to change (probably to reduce) the
      value of ϵ from that determined from these experiments.
      Note also that in view of possible uncertainty about the

value of h, there is an argument for analysing the data
on the basis of (2.3.3), viz. $d\bar{x}/dt = \bar{u}(\epsilon' \bar{z})$, rather
than (2.3.4).

(iii) As stated above, there is a strong case, as far as
spreading and entrainment are concerned, for ignoring
possible effects of wind shear in a first analysis of
the data. Only if existing models, such as DENZ, or
that due to Fay and Ranck, or models like that
summarized in Table 2 but with $\cos\phi = 1$ for all t, prove
unacceptable, should more complicated models like that
proposed earlier be used instead.

(iv) Finally, recalling that the sole aims of box models are
to predict reliably the gas concentration and cloud
position, it follows that variables like r and v are
subsidiary and are introduced only to achieve these
aims. There is therefore no a priori reason to suppose
that the theoretical values of r, h and v are equal to
the visible radius, height and (inferred) volume, or
indeed any other radius, height and volume that may be
calculated from the experimental readings of concentration (e.g. by numerical integration). Of course, all
the experimental evidence so far available shows, in
general, much better than order of magnitude agreement
between the theoretical values and those deduced (by
whatever means) from the data, and it is this agreement
that is one of the most important reasons for believing
that box models will prove to be practically reliable,
and robust, predictors. Nevertheless it seems clear
that all investigators who compare theory and experiment
should define precisely how the radius, height and
volume are to be determined from the experiments, and it
would be helpful if all investigators could use the same
precise definitions.

4. STRATIFICATION AND VARIABILITY

4.1 Stratification

The previous discussion has dealt exclusively with the situation when the ambient atmosphere is neutrally stratified.
Some box models, including DENZ (Fryer and Kaiser 1979) and
Eidsvik (1980), contain explicit rules to deal with situations
when the ambient atmosphere is not neutrally stratified. The
rule in DENZ is a prescription of the top entrainment velocity
u_t that depends on Pasquill stability category in a way that
uses experimental data of Taylor, Warner and Bacon (1970).
However these data were taken at heights much greater than

those relevant to the dispersion of heavy gas clouds, at least
in the gravity-dominated stage. Eidsvik's model includes
surface heat flux in the formula for u_t. Other box models like
that due to Fay and Ranck (1981) make no special allowance for
atmospheric stratification. Models based on numerical solutions of approximate partial differential equations and
discussed briefly in §1.2 include stratification effects by
modifying the eddy diffusivity in the way described on pps.
18-21 of Havens (1982). However these modifications are based
on the same data by Taylor, Warner and Bacon (1970) that are
used to modify DENZ, and whose relevance to heavy gas dispersion seems doubtful.

Since it is not yet true that available experimental data
have enabled one box model to be selected as the best, even for
releases into neutrally buoyant atmospheres, it is evident,
a fortiori, that modifications for atmospheric stratification
like those summarized above have not been validated against
experimental data. Unfortunately the same statement seems to
be true even for neutrally buoyant releases (i.e. for clouds
whose initial density is the same as that of the air at the
position of release).

One of the most important unsolved problems is how to
classify stratified atmospheres from the point of view of
practical efficacy. The most commonly used scheme is that in
terms of Pasquill stability category (Pasquill 1974) but, as
shown by Sedefian and Bennett (1980), it is often difficult to
use this scheme without ambiguity since, contrary to intention,
different measured atmospheric parameters often predict different stability categories. Since only releases of a limited
horizontal scale and time duration are of interest, a more
satisfactory classification from some points of view is in
terms of the Monin-Obukhov length D where (Monin and Yaglom
1971, Chapter 4)

$$D = - \frac{u_*^3 \, T_O}{\kappa g (\overline{w' T'})} .$$
(4.1.1)

In (4.1.1) T_O is the base temperature and $(\overline{w' T'})$ is the upwards
mean flux of temperature fluctuations. In stable conditions
(Pasquill categories E, F, G) D is positive, in unstable conditions (Pasquill categories A, B, C) D is negative and in
neutral conditions (Pasquill category D) $D = \infty$. In the Porton
experiments, the values of D were usually negative with a
modulus greater than 25m (Picknett 1978), while the initial

cloud height was 3.5m. For the Thorney Island trials h_O is
about 13m. However, as the result of the large initial
slumping, the condition h << D is likely to be satisfied for a
substantial period after release. It seems likely that, except
in very rare conditions, the same condition will hold in most
situations of practical interest (Chatwin 1981a, Fay and Ranck
1981). Under such circumstances it seems legitimate to ignore
atmospheric stability for a long time after release.

However, since box models all predict that eventually the
cloud height h increases as the result of ambient atmospheric
turbulence, atmospheric stratification will inevitably become a
significant influence on the dispersion. When h becomes
comparable with D, the models discussed earlier in this paper
will need (perhaps significant) modification. Consider, for
example, the box model of passive dispersion introduced in
§2.1. The entrainment velocities in equation (2.1.3), each
there proportional to u_* on dimensional grounds, must now be
equal to {u_* multiplied by an unknown function of (h/D)}. A
similar modification is necessary to (2.1.1). Discussion of
models of atmospheric dispersion which take stability into
account and reduce to the predictions of Lagrangian similarity
(and hence to the model in §2.1) when D = ∞ have been given by
Gifford (1962), Cermak (1963) and van Ulden (1978). Although
results appear to be consistent with Monin-Obukhov similarity
theory (provided the horizontal scale and duration of release
are not too large) there is not enough evidence to determine
the unknown functions of (h/D) that would need to be introduced
into the box models. For this reason, it is unrealistic to
attempt to quantify the effects of wind shear on entrainment in
circumstances when atmospheric stability is important. However,
wind shear effects also obey Monin-Obukhov similarity theory
and, in that sense, should be accountable for without intro-
ducing additional parameters.

The situation regarding advection is more satisfactory than
that for entrainment and spreading. In the presence of stratification the mean wind profile ū(z) is adequately approximated
by

$$\bar{u}(z) = \frac{u_*}{\kappa} \left[\ln\left(\frac{z}{z_O}\right) + \frac{\beta(z-z_O)}{D} \right] , \qquad (4.1.2)$$

(Monin and Yaglom 1971, p. 432) where β is a universal constant
whose value, though still uncertain, is about 5 (Turner 1979,
p. 131). Equation (2.3.2) with F(z) = 1 then gives

$$\frac{d\bar{x}}{dt} = \frac{u_*}{\kappa} \left[\ell n \left(\frac{h}{ez_0} \right) + \frac{\beta}{D} (\tfrac{1}{2}h - z_0) \right] .$$
(4.1.3)

Since $e^{-1} \neq \frac{1}{2}$, this is not accurately consistent with $d\bar{x}/dt = \bar{u}(\epsilon h)$, but the difference seems unlikely to be important, or unambiguously detectable in experiments.

4.2 Variability

In view of an earlier report (Chatwin 1981a) for HSE, it is appropriate to close this paper by repeating the opinion that, eventually, the assessment of potential hazards associated with dispersing heavy gas clouds will take explicit account of variability between experiments, variability arising from the inevitable turbulent fluctuations. Here, it is necessary only to point out (Chatwin 1981a, 1982) that estimates of fluctuations can be incorporated within the box model format, and that there is no reason to believe that the philosophy of box models does not apply to fluctuations of concentration, as it does to the ensemble mean.

5. ACKNOWLEDGEMENTS

The work described in this paper was supported by the U.K. Health and Safety Executive under Contract No. 1189.1/01.01. and I am grateful to Mr. R. Fitzpatrick of HSE for his efficient computing, and to Dr. J. McQuaid of HSE for his encouragement and practical wisdom. I was also helped by many other people especially Dr. P.W.M. Brighton, Dr. D.M. Webber and Dr. C.J. Wheatley of SRD, Dr. D.J. Hall of Warren Spring Laboratory and Dr. J.S. Puttock of Shell.

6. REFERENCES

Batchelor, G.K. (1964) Diffusion from sources in a turbulent boundary layer, *Arch. Mech. Stosowanej,* **16**, 661-670.

Britter, R.E. (1982) Special topics on dispersion of dense gases, Report on Contract No. 1200/01.01, Research and Laboratory Services Division, Health and Safety Executive, Sheffield, February 1982.

Cermak, J.E. (1963) Lagrangian similarity hypothesis applied to diffusion in turbulent shear flow, *J. Fluid Mech.,* **15**, 49-64.

Chatwin, P.C. (1968)   The dispersion of a puff of passive con-
    taminant in the constant stress region, *Q.J.R. Met. Soc.*,
    94, 350-360.

Chatwin, P.C. (1970)   The approach to normality of the concen-
    tration distribution of a solute in a solvent flowing along
    a straight pipe, *J. Fluid Mech.*, **43**, 321-352.

Chatwin, P.C. (1971)   On the interpretation of some longitu-
    dinal dispersion experiments, *J. Fluid Mech.*, **48**, 689-702.

Chatwin, P.C. (1981a)   The statistical description of the
    dispersion of heavy gas clouds, Report on Contract No.
    1189/01.01, Research and Laboratory Services Division,
    Health and Safety Executive, Sheffield, February, 1981.

Chatwin, P.C. (1981b)   The influence of basic physical
    processes on the statistical properties of dispersing heavy
    gas clouds, Proc. 7th Biennial Symposium on Turbulence,
    Rolla, Missouri.

Chatwin, P.C. (1982)   The use of statistics in describing and
    predicting the effects of dispersing gas clouds, *J. Haz.
    Mat.*, **6**, 213-230.

Chatwin, P.C. (1983)   The incorporation of wind shear effects
    into box models of heavy gas dispersion, Report on Contract
    No. 1189.1/01.01, Research and Laboratory Services Division,
    Health and Safety Executive, Sheffield, February 1983.

Chatwin, P.C. and Sullivan, P.J. (1981)   Diffusion from an
    elevated source within the atmospheric boundary layer, Proc.
    Third Int. Symp. on Turbulent Shear Flows, Univ. of
    California at Davis, 9.1-9.6.

Cox, R.A. and Carpenter, P.J. (1980)   Further development of a
    dense vapour cloud dispersion model for hazard analysis.   In
    "Heavy Gas and Risk Assessment" (S. Hartwig, Ed.), D. Reidel,
    Dordrecht, 55-87.

Eidsvik, K.J. (1980)   A model for heavy gas dispersion in the
    atmosphere, *Atm. Envir.*, **14**, 769-777.

Fay, J.A. (1982)   Some unresolved problems of LNG vapor
    dispersion.   In "Dispersion of dense vapors" (J.A. Havens,
    Ed.), Volume II of MIT-GRI LNG Safety and Research Workshop,
    Gas Research Institute, Chicago, 72-84.

Fay, J.A. and Ranck, D. (1981) Scale effects in liquified fuel
    vapor dispersion, DOE/EP-0032, U.S. Dept. of Energy,
    Washington, D.C., December 1981.

Fitzpatrick, R.D. (1981) A comparison of two solutions to the
    problem of predicting the dispersion of a gas cloud released
    to the atmosphere, section paper IR/L/HA/81/6, Health and
    Safety Executive, Sheffield.

Fryer, L.S. and Kaiser, G.D. (1979) DENZ - A Computer program
    for the calculation of the dispersion of dense toxic or
    explosive gases in the atmosphere, SRD R 152, Safety and
    Reliability Directorate, UKAEA, Culcheth, Warrington.

Gifford, F.A. (1962) Diffusion in the diabatic surface layer,
    J. Geophys. Res., **67**, 3207-3212.

Hall, D.J., Hollis, E.J. and Ishaq, H. (1982) A wind tunnel
    model of the Porton dense gas spill field trials, LR 394
    (AP), Warren Spring Laboratory, Stevenage.

Havens, J.A. (1982) (Ed.) Dispersion of dense vapors, volume
    II of MIT-GRI LNG Safety and Research Workshop, Gas Research
    Institute, Chicago.

Hunt, J.C.R. and Weber, A.H. (1979) A Lagrangian statistical
    analysis of diffusion from a gound-level source in a turbu-
    lent boundary layer, *Q.J.R. Met. Soc.*, **105**, 423-443.

McQuaid, J. (1982) Future directions of dense-gas dispersion
    research, *J. Haz. Mat.*, **6**, 231-247.

Monin, A.S. and Yaglom, A.M. (1971) Statistical Fluid
    Mechanics, Vol. 1, (J.L. Lumley, Ed.), The MIT Press.

Morton, B.R., Taylor, G.I. and Turner, J.S. (1956) Turbulent
    gravitational convection from maintained and instantaneous
    sources, *Proc. Roy. Soc. A,* **234**, 1-23.

Pasquill, F. (1974) Atmospheric Diffusion, Ellis Horwood.

Picknett, R.G. (1978) Field experiments on the behaviour of
    heavy gas clouds, Ptn. IL 1154/78/1, CDE, Porton Down, Wilts.

Raj, P.K. (1982) Heavy gas dispersion - a state-of-the-art
    review of the experimental results and models. Lectures
    presented at von Karman Institute for Fluid Dynamics,
    Rhode-Saint-Genese, Belgium, March 1982.

Sedefian, L. and Bennett, E. (1980) A comparison of turbulence classification schemes, *Atm. Envir.*, **14**, 741-750.

Taylor, R.J., Warner, J. and Bacon, N.E. (1970) Scale length in atmospheric turbulence as measured from an aircraft, *Q.J.R. Met. Soc.*, **96**, 750-755.

Turner, J.S. (1979) Buoyancy Effects in Fluids, Cambridge University Press.

van Ulden, A.P. (1974) On the spreading of a heavy gas released near the ground, Proc. First Int. Symp. on Loss Prevention and Safety Promotion in the Process Industries, Holland.

van Ulden, A.P. (1978) Simple estimates for vertical diffusion from sources near the gound, *Atm. Envir.*, **12**, 2125-2129.

Webber, D.M. (1983) The physics of heavy gas cloud dispersal, SRD R 243, Safety and Reliability Directorate, UKAEA, Culcheth, Warrington.

Woodward, J.L., Havens, J.A., McBride, W.C. and Taft, J.R. (1982) A comparison with experimental data of several models for dispersion of heavy gas clouds, *J. Haz. Mat.*, **6**, 161-180.

Yaglom, A.M. (1976) Diffusion of an impurity emanating from an instantaneous point source in a turbulent boundary layer, *Fluid Mechanics - Soviet Research*, **5**, 73-87.

MULTIPLE SMOKE PLUME TRIALS - THE CHEMICAL DEFENCE ESTABLISHMENT

C.D. Jones and D.J. Ride

(C.D.E. Porton Down)

A series of six trials was performed by CDE at RAE Cardington in July 1982 using the balloon facilities available there to photograph from aloft the plumes from a crosswind line of smoke grenades. Photographs were taken at half-second intervals over a period of up to one minute by a camera looking vertically down at a nominal height of 650ft or 1500ft depending on the length of the source line.

The aim of the experiments was to determine the feasibility of establishing a data bank for calculating the statistics of multiple plume behaviour. The trials were successful and the photographs are now being processed using a digitising tablet linked to a computer to yield plume centreline separation and velocity statistics. Some measurements of the mean width of the visible plume have been made and this analysis will be further explored using a computer-based image processor.

On-site measurements of radiation and wind components were made by the Meteorological Office during some of the trials, and these will enable the statistics to be related to particular levels of turbulence.

The plumes from a 100m line of grenades 15s after firing.
Balloon height 650ft. Note the grid of white marker cards.

The line of smoke grenades

The camera mounted on its boom.

The balloon and its winch.

VERTICAL HEAT TRANSPORT IN A COOLING WATER PLUME

J.H. Pickles and I.R. Rodgers

(CEGB, Central Electricity Research Laboratories, Leatherhead)

SUMMARY

   Predictions of cooling water dispersion from CEGB power stations are needed both for environmental reasons and in order to ensure minimum recirculation and maximum operating efficiency. This paper describes a series of measurements made at a coastal power station, with particular emphasis on the vertical temperature structure of the cooling water plume. This shows a well-defined warm surface layer whose thickness and temperature both decrease slowly as the plume moves down-stream with the tide. Heat flux values calculated from the observed temperature profiles clearly show the reduction in vertical transport produced by the stable stratification. For practical purposes a simple model is desired and some numerical experiments with an eddy diffusivity model have been done. This reproduces well the observed features of the plume decay although its quantitative predictions are sensitive to the input data.

1. INTRODUCTION

   Direct cooled power stations have a large requirement for condenser cooling water. For an electrical output of 1000 MW the flow taken from the sea or estuary approaches 50 m^3/s and this is discharged again with an excess temperature of about 10°C. The subsequent behaviour of this warm water discharge is of major importance in choosing the positions of the cooling water intake and outfall. Warm water recirculation leads to reduced thermal efficiency which must be balanced against the increased civil engineering costs associated, for example, with tunnelling to increase the separation of intake and outfall.

   Taken at the simplest level mathematical modelling provides two inputs to the design process. One, clearly, is a prediction

of the extent, in plan view, of the warm water plume. The
second is a prediction of the depth structure of the plume,
which controls the extent of the 'drawdown' of warm water into
the cooling water intake on the sea bed at those times when the
plume, or the pond which forms at slack water, passes over it.
Both predictions depend on understanding the nature of the
vertical mixing because it affects the vertical structure and
the depth of the plume. It also affects the gravity current
head at the edge of the plume whose spreading rate largely
determines the plume outline.

To back up these predictions several surveys at power
station sites have been done (Moore and James, 1973, Macqueen
and Howells, 1978, Macqueen and Parker, 1981) both from boats
and using infra-red photographs from aircraft - 'thermography'.
In the simplest case when the discharge is emitted into a tidal
stream running parallel to the shoreline, a more or less
parabolic profile is formed in plan. After a period of rapid
initial dilution, the surface temperature declines at a rate of
the order of 1°C/km as the plume moves downstream. These data
were in general insufficient to determine the mechanism of
temperature decay.

In 1980 and 1981 further plume surveys were made to give more
precise information on the variation of temperature with depth.
The site chosen was Sizewell, where the coastline is very nearly
straight and the power station discharges into a tidal flow with
an amplitude of 0.5 - 0.7 m/s. Water temperatures were
measured from a hired fishing vessel using thermistors fixed at
half metre intervals on a vertical spar. Temperatures,
salinities, depth sensor readings and signals from a microwave
'trisponder' position fixing system were recorded automatically
on a data logger at 5 second intervals as the boat zig-zagged
in and around the plume. Three recording current meters were
also deployed and some supplementary measurements were made
with a thermistor and a direct reading current meter from a
small inflatable dinghy.

2.   EXPERIMENTAL RESULTS

2.1.   *Temperature Data*

From the wide range of experimental measurements obtained we
select data which help to elucidate the physical processes
responsible for vertical heat transport when the plume is
discharged into a more or less steady tidal flow.

Fig. 1 shows the decay of surface temperature with distance
along the plume centre-line. All measurements were made in a
tidal stream whose recorded velocity was about 0.5 m/s; data

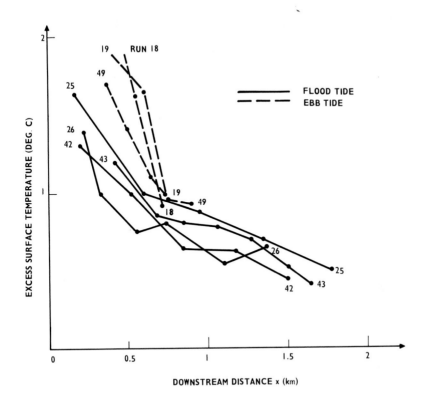

Fig. 1   Excess surface temperature for different plume runs

from runs 18, 19 and 49 refer to an ebb tide, with the plume
moving northwards from the outfall, while runs 25, 26, 42 and
43 were made on the flood tide. The measured temperatures show
a gradual decay with distance as seen at other sites.   An
unexpected feature is the clear difference between the plumes
on the ebb and flood tides, with ebb tide plumes showing
substantially higher temperatures. This may be caused by
differences in the amount of turbulence induced locally at the
outfall.

   Fig. 2 shows vertical temperature profiles in two of the
plumes. Again there are substantial differences between the
ebb tide (run 19) and flood tide (run 25) plumes. On the ebb
tide there is a broadly uniform vertical temperature gradient
of about 1°C/m down through the plume, but the flood tide plume
shows a 'core-and-gradient' structure, with a warm layer at a
uniform temperature above the region with a strong temperature
gradient.

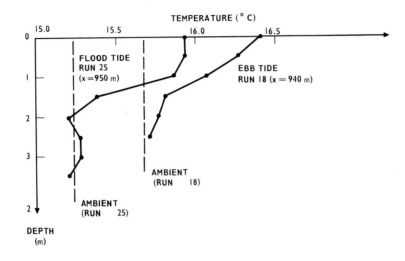

Fig. 2 Vertical temperature profiles through the plume

A further feature of the flood tide plumes is that their
cross-sectional width remains approximately constant from
distances of 300 m to 1500 m downstream of the outfall. This
appears to be the result of the ridge on the sea bed, shown in
Fig. 3, constraining the lateral spreading of the gravity
current at the edge of the plume. At all events, the constant
width plume provides a convenient natural test bed for analysis
of the vertical heat transfer. Fig. 4 shows the evolution of
the vertical temperature profile in the constant width plume as
it moves downstream.

2.2. Box Model Analysis

Box models, otherwise known as 'slab' or 'moving-slice'
models, are widely used to predict dispersion in stratified
flow (Van Ulden, 1974, Cox and Roe, 1977, Ewing, 1980). The
basic assumption is that the flow is separated into two clearly
distinguishable layers, with the density (or temperature)
profile approximated by a step function as shown in Fig. 5.
The further assumption of uniformity across the transverse
section of the plume is also reasonably well satisfied, at
least for moderate distances downstream.

The box model parameters $\bar{T}$ and $\bar{h}$ shown in Fig. 5 offer a
convenient data reduction scheme for the vertical temperature
distributions $T(z)$ measured in the flood tide plumes. For

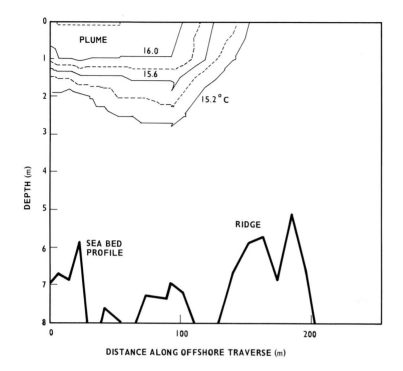

Fig. 3   Sea bed profile and plume temperature contours for
cross-section transverse to flow (Run 26)

plumes measured over the largest available depth range,
$0 \leqslant z \leqslant 3.5$ m, we set

$$\bar{T} = T(0.5) - T(3.5) \tag{1}$$

and

$$\bar{h} = \frac{\displaystyle\int_0^{3.5} [T(z) - T(3.5)]dz}{\bar{T}} \tag{2}$$

The equivalent core excess temperature $\bar{T}$ is defined by equation
(1) in such a way as to exclude both the effects of fluctuations in surface temperature $T(0)$ and also the gradual downstream warming of the ambient flow by the plume measured by the
bottom thermistor on the spar. The equivalent plume depth $\bar{h}$ is
defined by equation (2) and is evaluated by trapezoidal
integration over the data points at half-metre intervals.

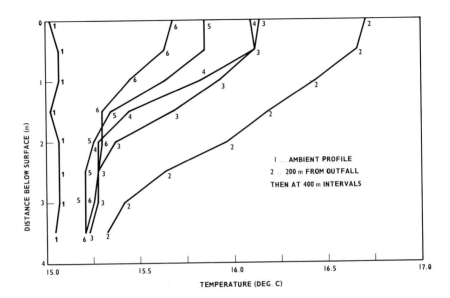

Fig. 4 Evolution of vertical temperature profile with distance
        downstream (Run 25)

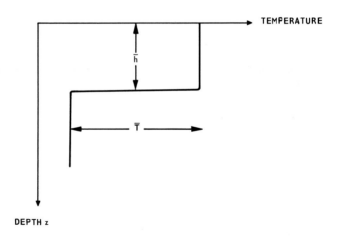

Fig. 5 Definition sketch for box model of a warm surface
        layer with parameters $\bar{T}$ and $\bar{h}$

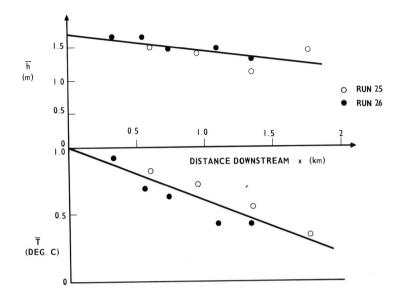

Fig. 6   Variation of box model parameters with distance
         downstream

The observed changes in $\bar{T}$ and $\bar{h}$ with downstream distance x
are plotted in Fig. 6.   Data from both run 25 (open circles)
and run 26 (full circles) have been included.   Both $\bar{T}$ and $\bar{h}$
decline steadily with distance, the linear regression lines
shown having slopes

$$\frac{d\bar{T}}{dx} = - 0.38^{\circ}C/km \qquad (3)$$

and

$$\frac{d\bar{h}}{dx} = - 0.26 \ m/km \qquad (4)$$

respectively over ranges of observations centred at $\bar{T} = 0.61^{\circ}C$
and $\bar{h} = 1.44$ m.

A further parameter of interest is the steepness of the
vertical temperature gradient dT/dz at the plume/ambient
interface.   Fig. 7 shows maximum gradients taken from cubic
functions interpolating the four data points T(0.5), T(1),
T(1.5), T(2) nearest the interface.   These are of order

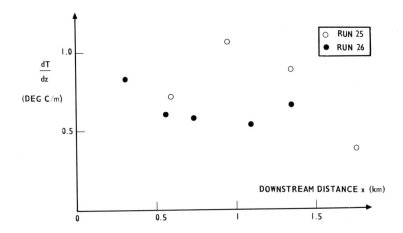

Fig. 7 Estimated maximum vertical temperature gradients at
        different downstream distances

$0.5-1^\circ$C/m and within the limits of the experimental scatter
there is no detectable change as the plume moves downstream.

2.3 Vertical Heat Transfer

The temperature data available also allow the estimation of
heat transfer rates. Suppose that $Q(x,z)$ denotes the vertical
heat flux at depth z and at a given downstream distance x.
Conservation of heat applied to a constant width plume of
uniform transverse cross-section in a tidal stream $u(z)$ leads
to the heat balance

$$\frac{\partial T}{\partial t} + u(z)\, \frac{\partial T}{\partial x} = -\,\frac{1}{\rho C}\, \frac{\partial Q}{\partial z} \tag{5}$$

where ρC is the specific heat of water per unit volume. In a
steady state plume $Q(x,z)$ can be estimated from measurements
$T(x,z)$ according to

$$Q(x,z) = -\,\rho C \int_0^z u(z)\, \frac{\partial T}{\partial x}\, dz \tag{6}$$

if it is assumed that there is no heat transfer across the sea
surface $z = 0$.

No detailed flow measurements were available and so for the
purposes of equation (6) $u(z)$ was approximated by the mean
flow u_c over the depth interval of interest $0 \leqslant z \leqslant 3.5$ m.
A self consistent value u_c was obtained by equating the total
heat flux in the plume to the total heat discharged from the
power station. In practice u_c was rather smaller than the
readings from the current meters positioned further offshore.
Fig. 8 shows heat fluxes Q calculated in this way for run 26.

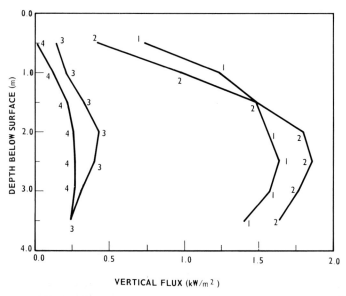

Fig. 8   The estimated heat flux $Q(x,z)$ as a function of depth
         for different downstream cross-sections x (Run 26).

The accuracy of this calculation is very sensitive to the
precision in the experimental measurements. The vertical heat
flux through the bottom of a typical control volume as sketched
in Fig. 9 lies in the range $0-2$ kW/m^2, or $0-10 \times 10^4$ kW integrated over the whole of the bottom surface. This is small
compared with total horizontal advective flows of order
5×10^5 kW, which could vary for example by $\pm 2 \times 10^4$ kW depending on small uncertainties ± 1 cm/s in the advective flow u_c, or
$\pm 0.04^\circ$C (half the temperature resolution of the data logger)
in the estimated ambient baseline temperature. Again, the
assumption in Equation (6) of zero heat transfer between the
atmosphere and the sea surface at the top of the control
volume is not exact. There is an approximate balance of the

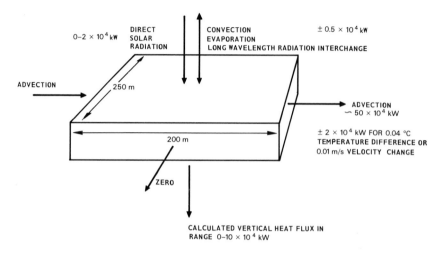

Fig. 9   Heat balance for typical control volume

conduction/convection/evaporation and long wavelength radiation
heat transfers, but direct solar radiation can also have
important effects (Kraus and Turner, 1967). Here the solar
input, though variable, reaches up to $0.5kW/m^2$ and for greater
precision this should also be allowed for.

3. EDDY DIFFUSION MODEL

We have sought to interpret the observations of section 2
using an eddy diffusion model of the kind described by Odd and
Rodger (1978). The heat flux Q is determined by an eddy
diffusivity Γ

$$Q = - \rho C \Gamma \frac{\partial T}{\partial z} \qquad (7)$$

where in an unstratified channel flow, with depth H and friction velocity u*, Γ has the standard parabolic profile

$$\Gamma_0(z) \sim u^*z \left(1 - \frac{z}{H}\right) . \qquad (8)$$

The effects of temperature stratification in water with thermal
expansion coefficient β are allowed for by including a further
factor f(Ri) which depends on the local gradient Richardson

number $Ri = \beta g |\partial T/\partial z| / [\partial u/\partial z]^2$. Thus

$$\Gamma(x,z) = f[Ri(x,z)]\Gamma_0(z) \tag{9}$$

and $f(Ri)$ takes the Munk-Anderson form (1948)

$$f(Ri) = (1 + aRi)^b = \left\{1 + \frac{a\beta g \left|\frac{\partial T}{\partial z}\right|}{\left(\frac{\partial u}{\partial z}\right)^2}\right\}^b \tag{10}$$

Current meter measurements taken from the dinghy (Fig. 10), though not sufficiently precise to allow detailed predictions, suggest that in the plume $\partial u/\partial z$ is somewhat greater than would be expected in the corresponding neutral flow. The product

$$\Lambda = a \left(\frac{\partial u}{\partial z}\right)^{-2}$$

which occurs on the right hand side of equation (10) has therefore been treated as an adjustable parameter. Despite the sensitivity of the heat balance described in Section 2.3, and some variation of the Γ-values estimated for different plume cross-sections, Rodgers (1983) then finds good agreement between experimentally determined values of Q derived via equations (7), (8) and (9) and values calculated using the form assumed in (10) for the suppression of mixing.

To test the model's ability to reproduce the general pattern of the plume's behaviour we have also run numerical simulations. We study the evolution of simple initial temperature distributions

$$T(z) = \frac{1}{2} A \left[1 + \tanh \frac{z-B}{C}\right] \tag{11}$$

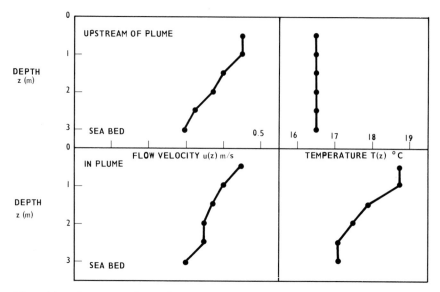

Fig. 10   Flow speeds and temperatures measured from the dinghy
          (runs 18/19)

under the diffusion process specified by equations (5) and (7).
Fig. 11 shows the result of such a simulation for conditions
approximating to those of runs 25 and 26, and a temperature
profile with A = 1°C, B = 1.5m, C = 0.5m. The constant of
proportionality in equation (8) was here taken as von Karman's
constant κ = 0.41, with a channel depth H = 7m, a friction
velocity u* = 0.02 m/s, and bulk tidal flow u_c = 0.3 m/s. The
adjustable parameter was set as Λ = 8.333 x $10^3 s^2$, which
corresponds to du/dz = 0.02 s^{-1} with Munk and Anderson's
recommended value a = 3.33.

    It can be seen in Fig. 11 that the initial temperature
distribution becomes rapidly steeper, reflecting the instability of the non-linear turbulent diffusion process when the
Richardson number becomes large (Puttock 1976, Posmentier
1977). In fact for the form

$$f(Ri) = \left\{ 1 + \beta g \Lambda \left| \frac{\partial T}{\partial z} \right| \right\}$$

it can be shown that the critical value of interfacial temperature gradient above which turbulent enhancement of gradients
occurs is

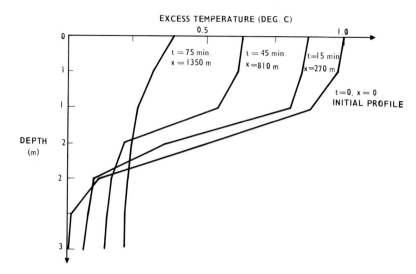

Fig. 11   Evolution of vertical temperature profile for eddy
diffusion model (H = 7m, u_c = O.3 m/s, u* = 0.02 m/s,
Λ = 8.33 x $10^3 s^2$)

$$\left.\frac{\partial T}{\partial z}\right|_{critical} = -\frac{1}{\beta g\Lambda (1+b)} \quad ,$$

neglecting the sensitivity of Λ to the value of $\partial T/\partial z$. Taking
b = $-^3/2$ and Λ = 8.333 x $10^3 s^2$ as specified above we find
$\left.\partial T/\partial z\right|_{critical}$ = O.12°C m^{-1}. The maximum plume gradients
plotted in Fig. 7 clearly exceed this critical value.   In the
numerical simulation the limiting gradient attained is fixed
by the vertical grid spacing Δ = 0.25 m and by the choice of
the numerical scheme, which is such that local values of heat
transfer are influenced by temperature values over a depth
interval 3Δ. Δ thus plays the role of a further adjustable
parameter in the model.   It provides a crude mathematical
representation of the physical smearing of the stable interface by the finite turbulent eddy size.

   Calculation of the subsequent development of the temperature profile shows the same features as those observed, viz:
cooling and erosion of the plume and a gradual warming

of the ambient flow beneath. We have not sought a formal fit
to the field data but it is useful to estimate the 'box' parameters $\bar{T}$ and $\bar{h}$ for comparison with those derived in Section
2.2. For the simulated plume we find

$$\frac{d\bar{T}}{dx} = - 0.59^{\circ}C/km$$

$$\frac{d\bar{h}}{dx} = - 0.26 \ m/km$$

over the range t = 15-60 min, x = 270-1080m, where $\bar{T}$ and $\bar{h}$ have
mean values $0.56^{\circ}C$ and 1.36m respectively. For the experimental
data $\bar{T}$ and $\bar{h}$ have mean values $0.72^{\circ}C$ and 1.52 m respectively
over this range of x values, and from equations (3) and (4)
$d\bar{T}/dx = - 0.38^{\circ}C/km$ and $d\bar{h}/dx = - 0.26 \ m/km$.

4.   DISCUSSION AND CONCLUSIONS

    There are many practical constraints in a full-scale survey
of a power station plume and it is difficult to make measurements as extensive or as precise as those which might be made
in a laboratory (Schiller and Sayre, 1975). Nevertheless the
results described here are sufficient to characterise the main
features of the vertical heat transport. These can be interpreted in terms of the eddy diffusion model of Section 3 which
ties in well with the observations without extensive 'tuning'.
Further work will be needed, however, for plumes which, like
the ebb tide plume of Fig. 2, do not have a box profile. The
model also needs to be generalized to the more common case when
the plume is not constrained to a constant width but spreads
horizontally under gravity (Larsen and Sorensen, 1968, Britter
and Simpson, 1978).

    A potentially more difficult problem is posed by the sensi-
tivity of the model to its input data. For example, Fig. 12
illustrates the variation of its predictions with changes in
the parameter $\Lambda = a(\partial u/\partial z)^{-2}$. The curves shown correspond,
with a = 3.33, to vertical velocity gradients $\partial u/\partial z = 0.15$,
0.20 and 0.25 s^{-1}. The velocity gradient is difficult to
measure, and perhaps even more difficult to predict, so that
predictions from the model will share this uncertainty. An
alternative approach to by-pass this problem might be to relate

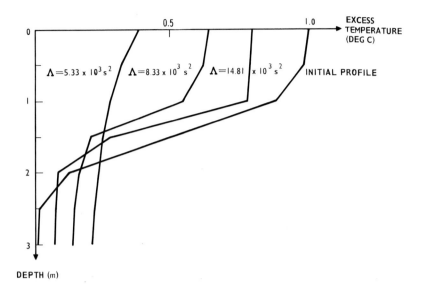

Fig. 12   Vertical temperature profiles at t = 45 min, x = 810m,
          for different values of eddy diffusion model input
          parameter

the rates of change of the box model parameters directly to
laboratory studies of the two-layer-mixing type (Turner 1973,
Thomas and Simpson 1983).   In any case, however, substantial
day-to-day variability in plume behaviour might be expected
from wind induced turbulence and solar radiation effects. This
question of the reliability and robustness of the theoretical
predictions deserves further consideration.

ACKNOWLEDGEMENT

The work described here was done at the Central Electricity
Research Laboratories at Leatherhead, Surrey, and is published
by permission of the Central Electricity Generating Board.

REFERENCES

Britter, R.E. and Simpson, J.E. (1978)   Experiments on the
    dynamics of a gravity current head, *J. Fluid. Mech.*, **88**,
    223-240.

Cox, R.A. and Roe, D.I. (1977)   A model of the dispersion of
    dense vapour clouds, 2nd International Symposium on Loss
    Prevention and Safety Promotion in the Process Industries,
    Heidelberg, September 1977.

Ewing, D.J.F. (1980) The vertical mixing of warm water plumes, Central Electricity Research Laboratories Report RD/L/N 35/80.

Kraus, E.B., and Turner, J.S. (1967) A one-dimensional model of the seasonal thermocline II. The general theory and its consequences, *Tellus*, **19**, 98-106.

Larsen, I. and Sorensen, J. (1968) Buoyancy spread of waste water in coastal regions, Proc. Coastal Engineering Conference, London.

Macqueen, J.F. and Howells, G.D. (1978) Waste heat dispersal - a cool look at warm water, CEGB Research, No. 7, 32-44.

Macqueen, J.F. and Parker G.C.C. Tidal currents measured near Sizewell Power Station, 1981, *Nucl. Energy,* **20**, 241-250.

Moore, D.J. and James, K.W. (1973) Water temperature surveys in the vicinity of power stations with special reference to infra-red techniques, *Water Research*, **7** (6), 807-820.

Munk, W.H. and Anderson, E.R. (1948) Notes on the theory of the thermocline, *J. Mar. Res.*, **1**, 276.

Odd, N.V.M. and Rodger, J.G. (1978) Vertical mixing in stratified tidal flows, Proc. ASCE, *J. Hydraulics Div.*, **104**, HY3, 337-351.

Posmentier, E.S. (1977) The generation of salinity fine structure by vertical diffusion, *J. Phys. Oceanography,* **7**, 298.

Puttock, J.S. (1976) Turbulent diffusion in separated and stratified flows, Ph.D Thesis, Cambridge University.

Rodgers, I.R. (1983) unpublished work.

Schiller, E.J. and Sayre, W.W. (1975) Vertical temperature profiles in open channel flow, Proc. ASCE, *J. Hydraulics Div.*, **101**, HY6, 749-761.

Thomas, N.H. and Simpson, J.E. (1983) Gravity currents in turbulent surroundings: laboratory studies and modelling implications, Proc. IMA Conference on Models of Turbulence and Diffusion in Stably Stratifed Regions of the Natural Environment, Cambridge, March 1983.

Turner, J.S. (1973) Buoyancy effects in fluids, Cambridge University Press.

Van Ulden, A.P. (1974) On the spreading of a heavy gas released near the ground, 1st International Symposium on Loss Prevention and Safety Promotion in the Process Industries.

INTENSE MIXING PERIODS IN AN ESTUARY

R.E. Lewis

(ICI PLC, Brixham Laboratory, Devon)

Intense mixing periods (IMPs) have been observed in the Tees estuary in north-east England during ebb tides. These have been detected by dye tracer studies and by observations of the vertical distribution of salinity. At about 2.5 hours after high water, the longitudinal concentration distribution at 0.5 m depth of a patch of dye (Fig. 1) changed from a spiky appearance (a) to a smoother form (b) within about 12 minutes. Fine structure reappeared when the detector was set to a higher sensitivity (c) but these oscillations were smoothed out in less than 10 minutes (d). Salinity profiles obtained in the same reach of the estuary showed a well stratified structure during the early part of the ebb (Fig. 2). By approximately 2.5 hours after high water (16.30), a distinct increase in the surface salinity had occurred and there was an appreciable decrease in the degree of stratification. These rapid changes in estuary structure have been ascribed to the onset of periods of intense mixing.

Abrupt increases in the salinity of the surface layer on an ebb tide have been observed in the Duwamish estuary, Washington State (Partch and Smith, 1978). Their observations indicated that there was a thickening of the surface layer which was too rapid to be explained by the downstream advection of the salt wedge. This suggested that intense mixing was taking place and that the major source of turbulent kinetic energy was in the upper layer. Partch and Smith suggested that the onset of an IMP was due to the breaking of long internal waves unable to travel upstream against the ebb current. The resulting internal hydraulic jump is responsible for the vigorous turbulent mixing. Internal waves are known to occur in stratified natural systems, carrying energy and momentum from regions of strongly turbulent flow (Townsend, 1976). In an estuary internal waves may be generated in the surface layer when water

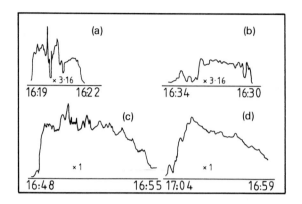

Fig. 1   Concentrations with time along a patch of fluorescent
         tracer dye, showing sensitivity ranges on the recording
         fluorometer.   Left hand sides of profiles correspond to
         the upstream edge of the patch.

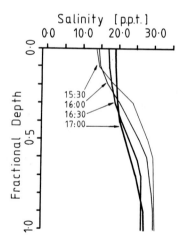

Fig. 2   Salinity profiles of station 6 in the Tees estuary on
         24th June 1980.   The total depth was approximately
         5 metres and highwater slack occurred at 14.00.

flows over a prominent topographic feature such as a sill or
ridge (Farmer and Smith, 1980). Although such waves are
capable of transferring momentum, the transfer of matter is
only possible when an instability occurs. The stability
criteria depend upon whether the flow can be regarded as a two-
layered system or a continuously stratified flow (Turner, 1973).
A Froude number for long internal waves can be expressed as,

$$F = u/(g\frac{\Delta\rho}{\rho}D)^{\frac{1}{2}}$$

where u is the mean flow velocity,
      $\Delta\rho$ is the top to bottom density difference
and   D is the depth of the upper layer which includes the
          thickness of the pycnocline

This internal Froude number is appropriate to a continuously
stratified system of depth D, and a critical value of $1/\pi$
for the Froude number has been predicted for such a flow (Long,
1955). However, if the surface waters are relatively well
mixed, then the system may be regarded as a two-layer flow and
the critical internal Froude number approaches unity, provided
that the surface layer thickness is small compared with the
total depth. Fig. 3 shows the change in Froude number plotted
relative to the time at which an IMP commences at three
positions in the Tees estuary, the distance between adjacent
sampling stations being approximately 2 kilometres. The Froude
numbers were computed from velocity and density profiles
obtained at half-hour intervals. Despite this limitation, the
plot indicates that the onset of intense mixing was associated
with a value for F of approximately 1.0, suggesting that the
system corresponded to a two-layer flow. Furthermore the IMPs
started at the upstream stations first and occurred progres-
sively later downstream. This feature could be a consequence
of the tapering cross-section of the Tees estuary towards the
landward end. As the ebb tide flow accelerates, the strongest
currents occur in the narrow reaches. Consequently, the Froude
numbers at the upstream stations become critical before
stations further to seaward. Alternatively, the changing start
times for the IMP along the estuary could be due to the advection of "broken water" downstream. If the internal waves break
high up the estuary shortly after high water, the ebb tide
would carry these mixed waters past the upstream stations first
and give the impression that the IMPs were commencing progres-
sively later downstream.

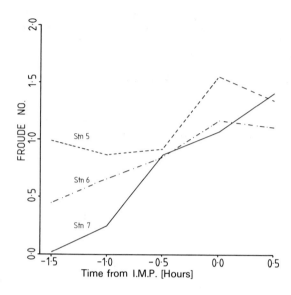

Fig. 3   Change in internal Froude numbers before and after the
start of the intense mixing periods at stations 5, 6
and 7.

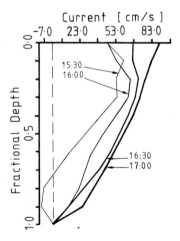

Fig. 4   Current profiles at station 6 on 24th June, 1980.

The significance of intense mixing periods in controlling the physical processes in an estuary can be seen from analyses of the tidal changes in estuary structure. Following an IMP, surface currents generally accelerate more rapidly than those in the lower layer (Fig. 4). This response could well be due to the breakdown of the internal waves and a readjustment of the shear stresses to more homogeneous conditions. IMPs also appear to be responsible for an effective phase shift between the velocity and salinity oscillations in an estuary, resulting in an upstream flux of salt which complements the flux due to the vertical circulation (Lewis and Lewis, 1983).

ACKNOWLEDGEMENT

The author wishes to express his thanks to Dr. A. New, Institute of Oceanographic Sciences, Taunton, for his comments on the draft manuscript.

REFERENCES

Farmer, D.M. and Smith, J.D. (1980) Tidal interaction of stratified flow with a sill in Knight Inlet, *Deep Sea Research*, **27A**, 239-254.

Lewis, R.E. and Lewis, J.O. (1983) The principal factors contributing to the flux of salt in a narrow, partially stratified estuary. *Estuarine, Coastal and Shelf Science*, **16**, 599-626.

Long, R.R. (1955) Some aspects of the flow of statified fluids. III. Continuous density gradients. *Tellus*, **7**, 341-357.

Partch, E.N. and Smith, J.D. (1978) Time dependent mixing in a salt wedge estuary. *Estuarine, Coastal and Shelf Science*, **6**, 3-19.

Townsend, A.A. (1976) The structure of turbulent shear flow. Second edition, section 8.16, Cambridge University Press.

Turner, J.S. (1973) Buoyancy effects in fluids. First edition, section 3.2, Cambridge University Press.